Introduction to Spectroscopy

Introduction to Spectroscopy

Dr Vijaykumar Deshmukh

ARCLER
PRESS

www.arclerpress.com

Introduction to Spectroscopy

Dr Vijaykumar Deshmukh

Arcler Press
4164 Lakeshore Road
Burlington, ON L7L 1A4
Canada
www.arclerpress.com
Email: orders@arclereducation.com

ISBN: 978-1-77956-958-5 (Hardcover)

ABOUT THE AUTHOR

Dr. Vijaykumar D Deshmukh is presently serving as Assistant Professor in Department of Plant Biotechnology, MGM College of Agricultural Biotechnology, Chh. Sambhajinagar (Maharashtra). He obtained his M.Sc. in (Agril. Biotech.) and Ph.D. in (Biotech.). He started his career as an Assistant Professor in 2005. His field of specialisation is Plant Biotechnology and Molecular Biology. He has published Six books and more than 25 research papers in National and international journals. He has participated in International, National and State Level Seminar and Symposiums.

TABLE OF CONTENTS

LIST OF FIGURES

PREFACE

Spectroscopy was devised in the 17th century following an experiment conducted by Isaac Newton, and the findings were published in 1672. Newton found in this experiment that sunlight has wavelengths spanning the whole visible spectrum, from roughly 390 nm to 780 nm, and all the colors of the rainbow. Indeed, he called this rainbow a spectrum. The discovery of new, invisible forms of electromagnetic radiation at the start of the 19th century, ultraviolet (UV) radiation by Ritter in 1801 at the short-wavelength and infrared (IR) radiation by Herschel in 1800 at the long-wavelength, expanded the spectrum offered by the Newton spectrum. These days, both spectral ranges are very important in various fields, including communications (IR) and environmental science (UV and IR). A potent experimental technique, nuclear magnetic resonance (NMR), can quickly and effectively handle a wide range of scientific activities at the molecular and atomic levels. This technique can be effectively used in the fields of physics, chemistry, biochemistry, geochemistry, biology, geology, pharmaceuticals, and materials science since an NMR signal can be detected in gases, solutions, liquids, and solids, including crystals and amorphous systems. Magnetic resonance imaging (MRI), a variation of NMR, has emerged as a diagnostic technique in veterinary and human science and medicine, especially for studies on the brain. The study of materials chemistry can also benefit from MRI. There is a vast array of NMR methods and techniques available today. Because the objects used in NMR research are becoming more sophisticated, this is always expanding and changing. Throughout nearly seven decades, NMR research has sparked the growth of new scientific disciplines. The first commercial NMR spectrometers appeared as a result of advancements in radiofrequency technology, which had a significant influence on chemistry, biochemistry, and biology research. The availability of potent NMR equipment with improved spectral resolution and sensitivity was necessary for the synthesis and investigation of complicated objects, such as biological molecular systems and intricate inorganic/organic molecular aggregates. Without advancements in magnet technology that produced stronger homogenous magnetic fields, such advancements would not have been feasible. Two, three, or four-dimensional NMR spectroscopy is now frequently used for structural investigations in solutions and solids to assign signals in the NMR spectra of complex chemical systems due to the development of computers that can process very rapid Fourier transforms (FFTs).

This book serves as a handbook for students interested in knowing the basics of spectroscopy. It can be used by students at all levels.

CHAPTER 1

BASICS OF SPECTROSCOPY

1.1. INTRODUCTION

Carbon and hydrogen are present in almost all organic molecules. Carbon must be present if combustion results in carbon dioxide, and hydrogen atoms must be present if combustion results in water (Robertson and Shaffer, 2014). Gas chromatographic techniques may identify water and carbon dioxide. After burning to produce nitrogen oxides, nitrogen atoms are frequently chemically reduced to nitrogen gas, whereas sulfur atoms are transformed into sulfur dioxide. When a chemical ignites in a hydrogen gas atmosphere, oxygen can be detected; water is the end product. Gas chromatography, which can also ascertain the relative concentrations of each of these gases, is currently used for such analyses (Novák, 1974). The computer can determine the sample composition as a percentage if the amount of the original sample is known. It is very uncommon to find a research laboratory where elemental analyses are carried out on-site, unless you work for a large corporation or one of the larger universities. Setting up the equipment and maintaining it within the limits of appropriate accuracy and precision takes time. Samples are typically sent to a commercial microanalytical laboratory that is equipped to perform this task on a regular basis and confirm the precision of the findings. Prior to the development of current equipment, the carefully weighed sample was burned in a glass tube that was cylindrical and housed inside a furnace. The unheated tubes containing chemical substances absorb carbon dioxide (NaOH/silica) and water ($MgClO_4$). A stream of oxygen was directed through the heated tube. The quantities of carbon dioxide and water formed were measured by removing and reweighing these detachable pre-weighed absorption tubes. Simple stoichiometry was used to determine the original sample carbon and hydrogen percentages (Chen et al., 2006).

The amount of oxygen was calculated by difference, which is standard procedure. One must ascertain the percentages of only C and H in a sample that only contains C, H, and O; oxygen is thought to be the non-calculated component. This technique can also be used when dealing with elements other than oxygen; if one of the elements is identified, it can be identified by difference. The computerized instrumentation performs the majority of calculations automatically. However, knowing the basic concepts behind the computations is frequently helpful for chemists. Mass spectrometry (Figure 1.1) is used in a modern laboratory to determine the molecular mass (McLafferty, 2011). The vapor density method is an obsolete technique that is, however, infrequently applied. This approach involves weighing a known volume of gas at a known temperature. We can calculate the mole fraction that the gas volume represents once it has been converted to standard temperature and pressure. We can quickly determine the molecular mass from that percentage. Measuring the freezing-point depression of a solvent caused by the addition of a known amount of the test substance is another way to calculate the molecular mass of a material. This approach is referred to as the cryoscopic method.

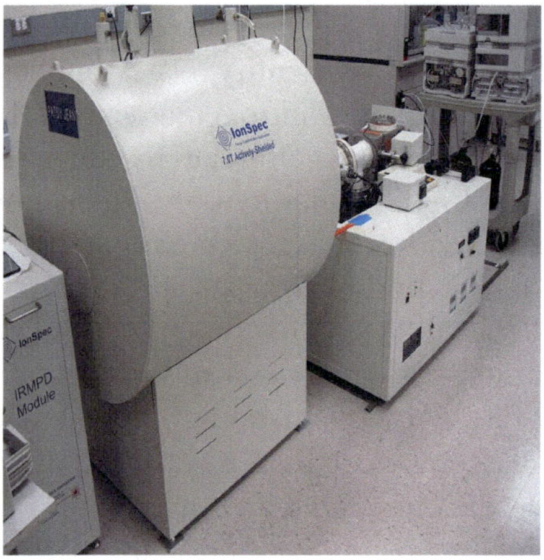

Figure 1.1. Mass spectrometry.

Source: https://upload.wikimedia.org/wikipedia/commons/a/ad/IonSpec_FT-ICR_%28Fourier_transform_Ion_cyclotron_resonance%29_Mass_spectrometer.jpg, Creative Commons License.

Vapor pressure osmometry is an additional technique that is rarely employed. It determines molecular weight by analyzing the change in a solvent's vapor pressure upon dissolution of a test substance in it. A standardized sodium hydroxide solution can be used to titrate the unknown material if it is a carboxylic acid. This process can be used to calculate a neutralization equivalent. The acid equivalent weight and the neutralization equivalent are the same. The molecular mass and neutralization equivalent are the same if the acid has only one carboxyl group (Leenheer et al., 2003). The neutralization equivalent of an acid with many carboxyl groups is calculated by dividing its molecular mass by the number of carboxyl groups. Many phenols, particularly those that have electron-withdrawing or electron-attracting groups, are acidic enough to be titrated using the same technique as sulfonic acids (Liu et al., 2015). The molecular mass and the empirical formula weight are frequently the same. In these situations, the molecular formula is also the empirical formula. Finding the number of times the empirical formula weight can be divided by the molecular mass is necessary since, in many situations, the empirical formula weight is less than the molecular mass. The component that must be multiplied by the empirical formula in order to obtain the molecular formula is determined in this way. Ethane is one example to describe this. The empirical formula for ethane, as determined via quantitative element analysis, is CH_3. The molecular mass is found to be 30. Ethane has an empirical formula weight of 15, which is equal to half of its molecular mass of 30. Ethane's chemical formula must therefore be $2(CH_3)$ or C_2H_6. In the infrared portion of the EMS, or electromagnetic spectrum, practically any substance with covalent bonds, whether organic or inorganic, absorbs different electromagnetic radiation frequencies. This region has wavelengths that are shorter than those linked to microwaves, but longer than those linked to visible light, which vary from around 400 to 800 nm. The vibrational component of the infrared spectrum is important. It includes light with wavelengths ranging from 2.5 μm to 25 μm (Campanella et al., 2021). You will frequently see the micron (μ) used on infrared spectra, even though the micrometer (μm) is the technically correct unit for wavelength in the infrared part of the spectrum.

1.2. SPECTRUM

The energy and frequency are directly proportional. The X-ray region of the EM spectrum, where the energy may be high enough to break molecular bonds, correlates to the highest energy radiation(Figure 1.2). Radiofrequencies at the other end of the electromagnetic spectrum contain extremely low energies, just sufficient to induce nuclear magnetic resonance (NMR) or electron spin resonance (ESR), which are nuclear or electronic spin transitions within molecules(Mallick, 2023). Instead of using wavelength (μ or μm), the majority of chemists use a unit called a wavenumber () to describe the radiation in the vibrational infrared portion of the electromagnetic spectrum. By calculating the reciprocal of the wavelength stated in centimeters, one may readily calculate wavenumbers, which are expressed as reciprocal centimeters (cm^{-1}). Multiply a wavenumber by the speed of light, which is measured in centimeters per second, to get a frequency v. Because wavenumbers are directly related to energy, a larger wavenumber denotes a higher energy, chemists prefer to use them as units. Therefore, the vibrational infrared ranges from roughly 4000 to 400 cm^{-1} in wave numbers(Efimov and Pogareva, 2006). Molecules that absorb infrared radiation are stimulated to a higher energy state, much like in other forms of energy absorption. Infrared radiation absorption is a quantized process, just like other absorption processes. Only specific infrared light frequencies (energies) are absorbed by molecules. Infrared light absorption results in energy shifts of between 8 and 40 kJ/mole(De Maleissye et al., 1976).

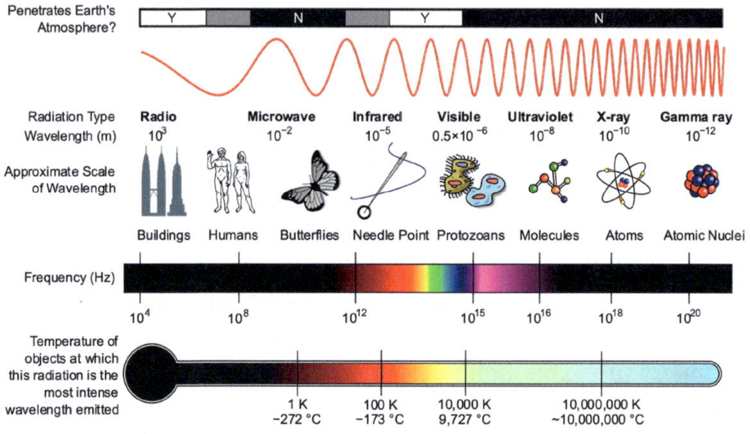

Figure 1.2. EM spectrum.

Source: https://upload.wikimedia.org/wikipedia/commons/c/cf/EM_Spectrum_Properties_edit.svg, Creative Commons License.

The range that includes the stretching and bending vibrational frequencies of the bonds in the majority of covalent compounds is represented by radiation in this energy range. Infrared radiation (IR) frequencies that correspond to the molecular vibrational frequencies are absorbed during the absorption process, and the energy absorbed helps to increase the amplitude of the vibrational motions of the molecular bonds. However, consider that not every bond in a molecule may absorb infrared radiation, even if the radiation frequency precisely overlaps with the motion of the bond. The ability to absorb infrared radiation is limited to bonds with a dipole moment that varies with time. H_2 and Cl_2 are examples of symmetric bonds that do not absorb infrared light. For energy to be exchanged, a bond needs to have an electrical dipole that is shifting at the same frequency as the incoming radiation. The incoming radiation (fluctuating electromagnetic field–sinusoidally) can then pair with the bond shifting electrical dipole(Barnes et al., 2018). As a result, a symmetric bond with almost or exactly the same groups on both ends won't absorb infrared light. The bonds of symmetric or pseudosymmetric alkenes () and alkynes () are the ones that are most likely to be impacted by this. No two molecules of different structures have the same IR or infrared absorption pattern, or infrared spectrum, because each type of bond has a unique natural frequency of vibration and two molecules of the same bond type are in slightly different locations. Two distinct molecules will never have the same infrared spectra, or patterns of absorption, even though some of the frequencies absorbed in the two scenarios may be the same. Therefore, just as a fingerprint may be used to identify a person, the infrared spectrum can be used to identify chemicals. You can determine whether two substances that are believed to be identical are actually identical by comparing their infrared spectra. Most of the time, two substances will be similar if their infrared spectra coincide peak for peak (absorption for absorption). Determining a molecule's structural information is a significant application of the infrared spectrum. Each bond type absorptions like $N-H, C \equiv C, C=C, C \equiv N$ and others are typically limited to specific areas of the vibrational infrared spectrum(Zhang et al., 2024). For any kind of bond, a narrow range of absorption can be identified. Absorptions outside of this range are typically caused by another kind of bond. For example, a $C-H$ bond in the molecule is usually responsible for any absorption in the 3,000 ± 150 cm^{-1} region, while a $C=O$ bond (carbonyl group) in the molecule is typically responsible for an absorption in the 1,715 ± 100 cm^{-1} range(Ahluwalia, 2023a). Every kind of bond has the same kind of range.

1.3. VIBRATION

Stretching and bending are the most basic forms, or modes, of vibrational motion in molecules that are infrared active, those that result in absorptions(Sinha et al., 2023). But there are also intricate forms of bending and stretching. Stretching vibrations often occur at higher frequencies than bending vibrations, and asymmetric stretching vibrations occur at higher frequencies than symmetric stretching vibrations. The initial infrared bands are frequently described in the literature using the terms scissoring, wagging, rocking, and twisting. There are two types of stretching in any group of three or more atoms, with at least two of them being identical: symmetric and asymmetric. Anhydrides, $-NO_2$, $-NH_2$, $-CH_3$, and $-CH_2-$ are a few examples of these groups. A symmetric stretching vibration at approximately 2,872 cm^{-1} and an asymmetric stretch at around 2,962 cm^{-1} are produced by the methyl group(Yusuf, 2023). The asymmetric and symmetric stretching modes of the anhydride functional group result in two absorptions in the $C = O$ region. The amino group exhibits a similar behavior, with a secondary amine (R_2NH) having only one absorption peak and a primary amine (NH_2). Amides have similar bands. For a nitro group, there are two prominent $N = O$ stretch peaks: the symmetric stretch is located at roughly 1350 cm^{-1}, and the asymmetric stretch is located at roughly 1,550 cm^{-1}(Srividya et al., 2022). We consider fundamental absorptions, which are nothing but vibrations. They result from the ground state being stimulated to the excited state with the lowest energy.

The existence of weak overtone and difference bands typically results in a complex spectrum. Excitation from the ground state to higher energy states, which correlate to integral multiples of frequency (n), produces overtones. Weak overtone bands, for instance, could be seen at $2\overline{v}, 3\overline{v}$ etc. Overtones are produced by any type of physical vibration. A fundamental frequency is produced when a string is plucked on a cello. However, a number of overtone frequencies are also used to set much less powerful vibrations. An overtone is a lower-intensity peak that may follow an infrared absorption at 500 cm^{-1} at 1,000 cm^{-1}(Motahari et al., 2023). A combination band is created when two vibrational frequencies of a molecule pair to produce a new frequency vibration inside the molecule. This vibration is infrared active. The two interacting bands add up to this band. Not every potential combination happens. Combination bands and difference bands are similar. The difference between the two interacting bands yields the observed frequency. By directly adjusting frequencies in wavenumbers by addition, subtraction, and multiplication, one can compute overtone, combination, and difference bands, respectively. Fermi resonance is the term used to describe the coupled vibration that results from a fundamental vibration coupling with an overtone or combination band. Only specific combinations are permitted. It is common to see Fermi resonance in carbonyl compounds. Even while the entire molecular rotational frequencies are not infrared active, they frequently combine with the molecular bending and stretching vibrations to give these absorptions fine structure, which further complicates the spectrum. A significant amount of unresolved fine structure may result from rotational coupling, which is one of the reasons a band in the infrared spectrum is broad rather than sharp.

Diatomic molecule or a molecule having two atoms, is two vibrating masses joined by a spring. Although the bond distance is always changing, it is possible to define an average bond distance. The potential energy rises whenever the spring is compressed or stretched past this equilibrium distance. When a bond vibrates, its energy of vibration alternates periodically and continuously between kinetic and potential energy, much like any other harmonic oscillator. A bond strength, length, and force constant K areimpacted by resonance. Therefore, a ketone conjugated with a $C = C$ double bond absorbs at a lower frequency, close to 1,675 to 1,680 cm^{-1}, whereas a typical ketone has its $C = O$ stretching vibration at 1,715 cm^{-1}(Ahluwalia, 2023b). Resonance gives the $C = 0$ bond better single-bond character and lengthens its bond distance. An infrared spectrometer, or a spectrophotometer, is the device that measures a compound's absorption spectrum(Figure 1.3). The organic chemistry laboratory frequently uses two kinds of infrared spectrometers: dispersive and Fourier transform (FT) devices. Spectra of compounds in the range of 4,000 to 400 cm^{-1} are provided by both of these kinds of equipment(Ahmad and Ayub, 2022).

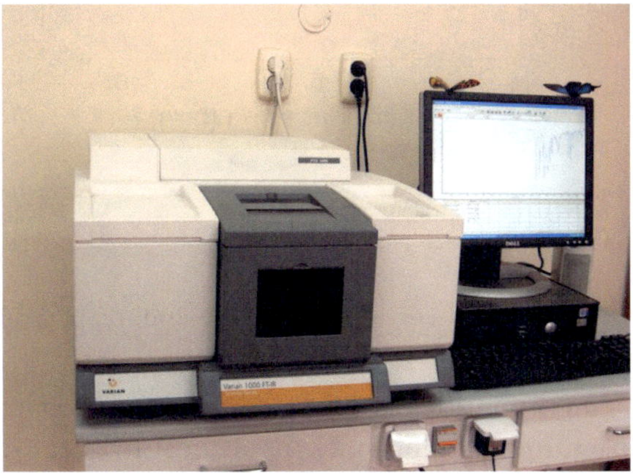

Figure 1.3. Infrared spectrometer.

Source: https://upload.wikimedia.org/wikipedia/commons/7/78/IR_spectrometer.jpg, Creative Commons License.

The two beams arc altcrnatcly passed to a diffraction gratingvia a beam chopper, that makes up the monochromator. The frequency or wavelength of the radiation that reaches the thermocouple detector is changed by the slowly spinning diffraction grating. The ratio of the reference and sample beam intensities is detected by the detector. In this way, the detector can identify which frequencies are unaffected by the light passing through the sample and which frequencies have been absorbed by it. The recorder displays the spectrum on a chart following the amplification of the detector signal. It is crucial to understand that the diffraction grating rotates to change the frequency of infrared radiation, recording the spectrum as it does. A spectrum in the frequency domain is said to be recorded by dispersive devices. The advanced spectrophotometers, or infrared spectrometers, work on a different principle. An interferogram is a pattern created by the optical pathway design. Despite being a complicated signal, the interferogram wave-like pattern includes all of the infrared spectrum frequencies.

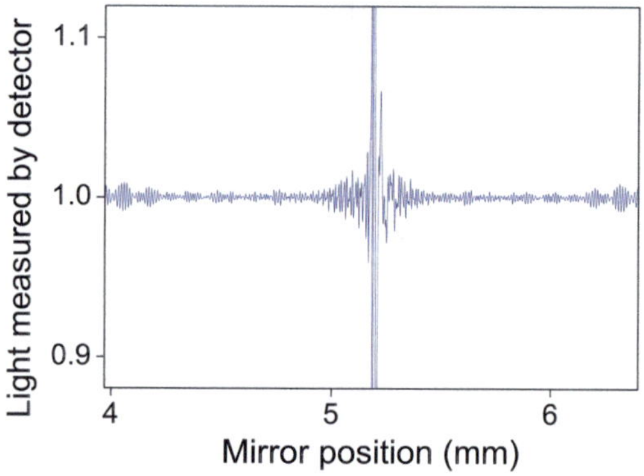

Figure 1.4. Interferogram.

Source: https://upload.wikimedia.org/wikipedia/commons/d/d4/FTIR-interferogram.svg, Creative Commons License.

An interferogram (Figure 1.4)is a time-domain spectrum that plots intensity against time. A spectrum that plots intensity against frequency, or a frequency-domain spectrum, is of interest to chemists. The interferogram distinct absorption frequencies can be separated using a mathematical process called a Fourier transform (FT), which results in a spectrum that is nearly identical to that measured by a dispersive spectrometer(Delsuc and O'Connor, 2024). This is called as Fourier transform infrared spectrometer, or FT-IR. An FT-IR instrument can acquire the interferogram in less than a second. Dozens of interferograms of the same sample can be collected and stored in a computer memory. An improved signal-to-noise ratio spectrum can be plotted by applying a Fourier transform to the total of the acquired interferograms. Therefore, an FT-IR instrument can be faster and more sensitive than a dispersion instrument.

1.4. INTERFEROMETER

The energy transmitted to the sample is processed by the FT-IR using an interferometer. A beam splitter, which is a mirror positioned at an $45°$ angle to the incoming radiation, allows the incoming radiation to pass through the interferometer but divides it into two perpendicular beams, one of which is undeflected and the other of which is orientated at a 90° angle. One beam travels to a stationary mirror before returning to the beam splitter. The beam splitter receives the undeflected beam after passing via a moving mirror. The pathlength that the second beam travels vary depending on how the mirror moves. Both constructive and destructive interferences are caused by the pathlength differences of the two beams, which recombine when they join at the beam splitter. Interferogram refers to the combined beam that contains these patterns (interference). This interferogram spans a broad range of wavelengths and includes all of the radiation originating from the source. The beam splitter directs the interferogram that is produced by merging the two beams toward the sample. All of the wavelengths, or frequencies, that are typically present in its infrared spectrum are absorbed by the sample as it travels through it. The computer compares the modified interferogram to a reference laser beam to establish a standard of comparison. The modified interferogram signal that reaches the detector contains information about the quantity of energy that was absorbed at each wavelength (frequency). All of the data is combined into a single time-domain signal in the final interferogram. To extract the various frequencies that were absorbed and to recreate and visualize what we know to be a normal infrared spectrum, a computer must perform a process known as a Fourier transform. The basic ideas go back to the late 1890s, when Wien investigated the magnetic deflection of anode rays and concluded that they were positively charged, and J. J. Thomson established the mass-to-charge ratio of the electron(Giliberti and Lovisetti, 2024). J. J. Thomson investigated the mass spectra of atmospheric gases and proved that elements might have isotopes by using a mass spectrum to show that Neon-22 existed in a sample of Neon-20. The first mass spectrometer was constructed by A. J. Dempster. Mass spectrometry was introduced in the 1980s and 1990s as a result of the development of sample introduction and ionization procedures for biological samples and high molecular weight (MW) chemicals. Mass spectrometry has become an essential technology in many sectors. Proteins, oligonucleotides, and polysaccharides are now assessed and sequenced in the biotechnology sector using mass spectrometry. Mass spectrometry is used by the pharmaceutical industry at every stage of the drug development process, including pharmacokinetics and drug metabolism, synthetic and combinatorial chemistry,compound discovery, and structural analysis.

1.5. MASS SPECTROMETRY

Mass spectrometry is used in medical field all over the world to test blood and urine for a variety of conditions, including the presence and concentrations of specific compounds that serve as markers for various disease states, such as types of cancer, as well as the detection and quantitative analysis of illegal drugs. Mass spectrometry is used by geologists to assess the quality of petroleum reserves and by environmental scientists to track the quality of the air and water. In order to find explosive remnants, mass spectrometry is also frequently employed in forensic investigations and airport security screening. The sample molecules are converted into gas phase ions at the ion source, which is from the sample intake. Recently, certain devices have been created that integrate the ion source and sample inlet in ambient settings, significantly streamlining sample preparation. An electromagnetic field is then used to accelerate the ions. The sample ions are then separated by the mass analyzer (mass-to-charge (m/z) ratio). The data system, usually a personal computer (PC), records and processes the signal after the detector counts the ions. The mass spectrum, a graph showing the quantity of ions found in relation to their m/z ratio, is the data output. The mass spectrometer is more complicated than is initially thought when we look closely at each of these crucial functions. A stream of molecules must be added to the ion source, also known as the ionization chamber, where the ionization occurs, before the ions may form. This stream of molecules is supplied by a sample intake mechanism. Mass spectrometry can analyze samples that are solid, liquid, or gaseous. To create the stream of molecules that must enter the ionization chamber, a sufficient amount of the sample must be transformed into the vapor state(da Silva Sousa et al., 2021). Since the material is naturally evaporated in gases, an inlet mechanism can be employed. The ionization chamber itself is at a lower pressure than the sample input system since this inlet system is only partially emptied or evaluated. In order to attract the vapor molecules into the low-pressure ionization chamber, the sample is placed into a bigger reservoir. Before entering the ionization chamber, the vapor passes through a tiny opening known as a molecular leak to guarantee that a constant flow of molecules is entering the chamber. Both volatile liquids and solids can be handled using the same technology. The apparatus can be made to fit within an oven, which can heat the sample to raise its vapor pressure for less volatile molecules. It is possible to heat the probe, which will evolve sample vapor close to the electron ionizing beam. At room temperature, molecules with vapor pressures less than 10^{-9} mmHg can be studied using a system like this.

1.6. CHROMATOGRAPH

A chromatograph is connected to the mass spectrometer to create the flexible inlet systems (sample)(Pastor et al., 2023).Using sample introduction method, the chromatograph separate components, allowing determination of each mass spectrum. One disadvantage of this approach is that the mass spectrometer must scan quickly. To prevent the first material from becoming contaminated by the next before its spectrum has been determined, the device must ascertain the mass spectrum of each component in the combination before the subsequent component leaves the chromatography column. Generally, chemicals are fully separated before the eluent stream is examined since the chromatograph uses high-efficiency columns. Over the m/z range, the device must be able to obtain at least one scan each second. If a smaller range of masses is to be examined, even more scans are required. High resolution and compactness are requirements for the mass spectrometer connected to the chromatograph. The gas stream that emerges from a gas chromatograph is entered through a valve into a tube, where it crosses a molecular leak, in gas chromatography–mass spectrometry (GC-MS)(Figure 1.5).

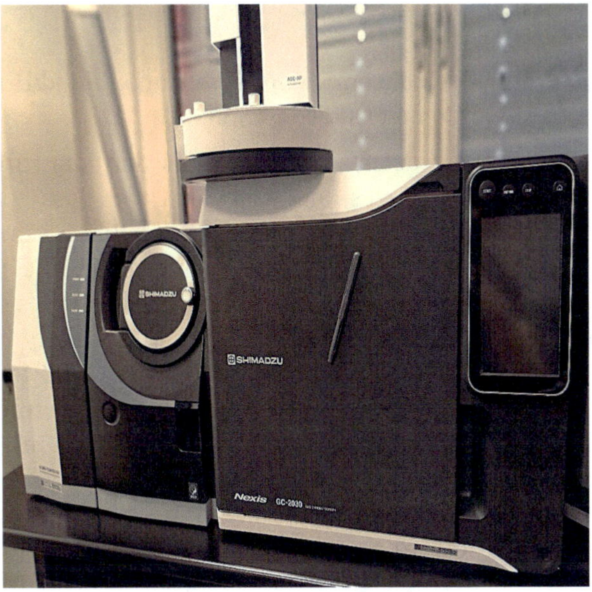

Figure 1.5. Gas chromatography–mass spectrometry.

Source: https://upload.wikimedia.org/wikipedia/commons/c/cd/GCMS_Instrument.jpg, Creative Commons License.

A portion of the gas stream is allowed into the mass spectrometer ionization chamber. The mass spectrum of each component in a mixture put into the gas chromatograph can therefore be obtained. The mass spectrometer essentially serves as a detector. High-performance liquid chromatography–mass spectrometry (HPLC-MS, or LC-MS)(Figure 1.6)uses an interface to connect an HPLC device to a mass spectrometer(Hemida et al., 2021).

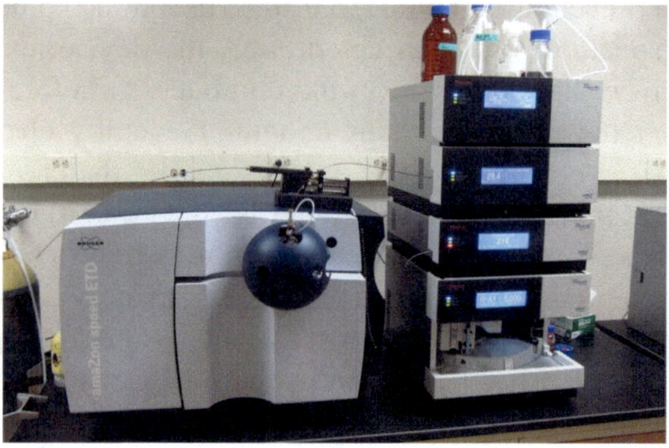

Figure 1.6. HPLC-MS.

Source: https://upload.wikimedia.org/wikipedia/commons/b/bd/Bruker_Amazon_Speed_ETD.jpg, Creative Commons License.

The mass spectrometer detects the chemicals that elute from the HPLC column, and it displays, analyzes, and compares the mass spectra of those substances with reference spectra from the computer library that is integrated into the device. Regardless of how the sample is introduced, the ion source must transform the stream of sample molecules into charged particles once they enter the mass spectrometer. Electron ionization (EI) is the widely used technique for turning the sample into ions(Famiglini et al., 2021). In EI-MS, a filament heated to1000°C or more emits a beam of high-energy electrons. The stream of molecules introduced from the sample inlet system is bombarded by these high-energy electrons. A cation is produced when an electron is removed from the molecule by an electron-molecule collision or bombardment. The freshly formed ions are guided toward a sequence of accelerating plates by a repeller plate, which has a positive electrical potential. A beam of quickly moving positive ions is created when a significant potential difference, between 1 and 10 kilovolts (kV), is delivered across these accelerating plates. The ions are directed into a uniform beam by one or more focusing slits. Vacuum pumps attached to the ionization chamber continuously attract the sample molecules, which are not ionized. By absorbing electrons, some of the molecules are changed into negative ions. These negative ions are absorbed by the Repeller plate. Certain instruments have the ability to flip the polarity of the accelerating and Repeller plates, enabling mass measurement of the negative ions (anions) produced by electron capture when the electron beam strikes the sample molecules. A small percentage of the generated positive ions might be charged more than once, meaning they have lost more than one electron. They accelerate similarly to positively charged ions that are singly charged (charged only once). Ionization potential, also known as ionization energy, is the energy needed to extract an electron from an atom or molecule. The ionization potentials of the majority of organic compounds range from 8 to 15 electron volts (eV)(El Ghazali et al., 2023). However, until an electron beam hits a stream of molecules with a potential of 50 to 70 eV, it cannot produce ions with high efficiency. A typical 70-eV electron beam is utilized to obtain spectrum properties, such as fragmentation patterns, that are easily comparable with electronic databases. For routine mass spectrometry of small organic compounds, EI-MS offers advantages. Hardware for electron ionization is reliable and reasonably priced or economical. The molecular ion is significantly fragmented as a result of the excess kinetic energy that the EI process imparts to the sample. There are many libraries of EI-MS data accessible, and a compound's fragmentation pattern can be replicated. This makes it easier to identify or validate a compound by enabling a PC to quickly compare the mass spectrum of a sample compound against millions of data sets in a spectral library. Another drawback could be the molecular ion fragmentation under EI conditions.

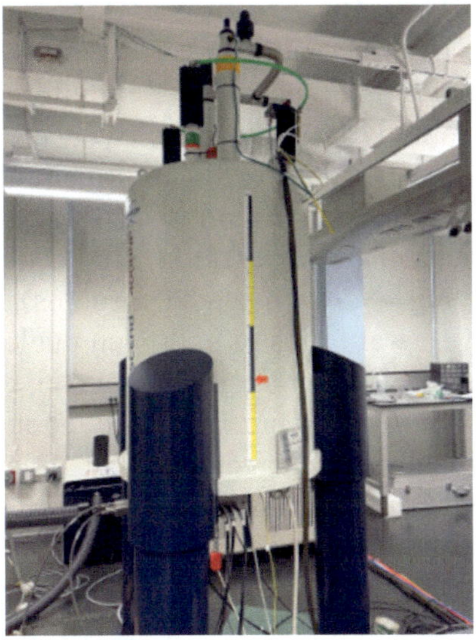

Figure 1.7. NMR.

Source: https://upload.wikimedia.org/wikipedia/commons/1/13/NMR_instrument.jpg, Creative Commons License.

Certain substances break up so quickly that the molecular ion lifetime is too short for the mass analyzer to identify them. Therefore, in these situations, it is impossible to calculate a compound's molecular mass. The requirement that the sample be relatively volatile in order to come into contact with the electron beam in the ionization chamber is another disadvantage of EI-MS. High molecular weight (MW) chemicals and the biomolecules are challenging to evaluate with EI-MS due to this fact and the fragmentation issue. One can alter the degree of ion fragmentation and the selectivity of ionization in CI-MS by changing the reagent gas. To guarantee effective ionization of the sample without excessive fragmentation, the reagent gas selection should be carefully considered in order to best match the reagent gas proton affinity with that of the sample. Energy is delivered to the sample during ionization in proportion to the difference between the reagent gas and the sample proton affinities. An analyte ion in a highly excited vibrational state is created by the excess energy. The sample ion will break due to the rupture of covalent bonds if excess kinetic energy is transmitted. As a result, employing a reagent gas whose proton affinity nearly matches that of the sample will produce more intact molecular ions and less fragment ions.

1.7. NUCLEAR MAGNETIC RESONANCE(NMR)

For organic chemists, nuclear magnetic resonance (NMR)(Figure 1.7)is an even more crucial spectroscopic technique than infrared spectroscopy. NMR techniques can be used to study a variety of nuclei, but the most widely accessible are hydrogen and carbon. NMR provides information about the quantity of magnetically different atoms, whereas infrared (IR) spectroscopy identifies the kinds of functional groups that are present in a molecule(Koczoń et al., 2023). For example, by studying hydrogen nuclei (protons), one can ascertain the quantity of each of the several forms of hydrogen nuclei and learn more about the characteristics of each type. For the carbon nuclei, comparable data can be found. IR and NMR data together are frequently enough to fully identify an unknown molecular structure.If a magnetic field of around 1.41 Tesla is applied to a proton, the energy difference between the proton two or double spin states is roughly 2.39×10^{-5} kJ/mole. This difference (energy) is associated with EMR radiation that has a frequency of 60 MHz, which is the radiofrequency (RF) part of the electromagnetic spectrum (EMS). Compared to hydrogen nuclei, other nuclei exhibit both larger and smaller energy disparities between spin states. The first nuclear magnetic resonance spectrometers provided a steady 60 MHz radiofrequency radiation and applied a variable magnetic field with a range of strengths close to 1.41 Tesla(Ritgen, 2025). They were ineffective for other nuclei and only successfully caused transitions between the proton (hydrogen) spin states in a molecule. To observe changes in the nucleus of other elements, like carbon and phosphorus, separate devices were needed. Fourier transform instruments are capable of observing the nucleus of multiple distinct elements within a single sensor. 300 MHz and 400 MHz equipment are rather common now, and major research centers have instrumentation with frequencies higher than 600 MHz. When a magnetic field is provided, protons start to precess, which causes them to absorb energy.

Precession is comparable to the spinning top. The top starts to wobble or precess, about its axis due to the gravitational field. When a magnetic field is supplied, a spinning nucleus exhibits comparable behavior. The nucleus starts to precess about its own axis of spin with an angular frequency, also known as its Larmor frequency, when a magnetic field is applied. The strength of the applied magnetic field directly affects the frequency at which a proton precesses; the higher the applied field, the higher the rate of precession. The frequency of precession for a proton is roughly 60 MHz if the applied field is 1.41 Tesla(Möbius and Savitsky, 2023). The precession creates an oscillating electric field with the same frequency since the nucleus is charged. The energy can be absorbed if the precessing proton is exposed to radiofrequency waves of this frequency. The two fields may couple and energy may be shifted from the incoming radiation (light) to the nucleus, resulting in a spin change, when the frequency of the oscillating electric field of the incoming radiation precisely matches the frequency of the electric field produced by the precessing nucleus. The nucleus is considered to be in resonance with the incoming electromagnetic wave when this state is present. Resonance occurs at around 60 MHz for a proton with an applied magnetic field of around 1.41 Tesla(Nimerovsky et al., 2021). Using $\Delta E = hv$, we can determine that the energy difference between the proton's two spin states is roughly 2.39×10^{-5} kJ/mole. Both of these energy levels can be filled with thermal energy from room temperature because there is slight energy separation between them. However, the lower-energy spin state has a surplus of nuclei. The Boltzmann distribution equations can be used to determine the size of this difference. We are able to observe resonance because of the extra nuclei. In addition to causing upward transitions, the application of 60 MHz radiation also stimulates downward transitions. We see no net signal if the upper and lower states are precisely equal. We can refer to this state as saturation. Saturation must be avoided when doing an NMR experiment. If the radiofrequency signal power is too strong, saturation occurs rapidly. As a result, NMR spectroscopy depends heavily on the extremely less nuclei in the lower spin state, and it is evident that highly sensitive NMR equipment is needed to pick up the signal. This excess increases when the NMR instrument's operating frequency rises because the energy difference between the two states expands. The excess rises with operation frequency. It also demonstrates why increasingly higher operating frequencies have been incorporated into the design of modern instruments. Since more nuclei can transition at higher frequencies, the instrument's sensitivity is raised and the resonance signals are stronger.

Since not every proton in a molecule has resonance at the same frequency, nuclear magnetic resonance is very useful. Molecular protons are surrounded by electrons and live in somewhat diverse electronic (magnetic) settings, which accounts for this variability. Each proton has a different valence-shell electron density. The surrounding electrons act as a shield for the protons. The proton valence electrons are moved when a magnetic field is introduced. A counter magnetic field is created by this circulation, also known as a local diamagnetic current, and it opposes the applied magnetic field. A magnetic field is created when electricity flows through a wire. Each proton in a molecule is protected from the applied magnetic field by diamagnetic anisotropy, which is dependent on the electron density around it. The induced counter field that opposes the applied field increases with the electron density surrounding a nucleus. The net applied magnetic field that a nucleus experiences is reduced by the counter field that surrounds it. The nucleus precesses at a lower frequency as a result. This indicates that radiofrequency radiation at this lower frequency is also absorbed by it. A molecule's resonance frequency varies slightly because each proton has a slightly different amount of electrical shielding due to its slightly different chemical environment. These resonance frequency variations are negligible. When the applied field is 1.41 Tesla, the resonance frequencies of the protons in fluoromethane and chloromethane differ by only 72 Hz(Gupta et al., 2023). The difference between fluoromethane and chloromethane indicates a frequency change of just one part per million because the radiation utilized to produce proton spin transitions at that magnetic field intensity has a frequency close to 60 MHz. Since it is extremely challenging to measure precise frequencies with that level of accuracy, limited effort is made to determine the precise resonance frequency of any proton. The resonance frequency of a proton in the sample is measured in relation to the resonance frequency of the protons of the reference material after a reference compound is added to the solution of the material to be analyzed. The frequency difference is immediately measured.

1.8. TETRAMETHYLSILANE (TMS)

Tetramethylsilane, or TMS is the standard reference chemical. Since the protons of its methyl groups are better protected than those of the majority of other known compounds, this molecule was initially selected. Since TMS would mark one end of the range, it was thought to be a good reference substance. At the time, no compounds were known to have better-shielded hydrogens than TMS(Ibrahim and Abdellatif, 2021). As a result, the resonances of the protons in another compound are measured and reported in terms of the Hertz from those of TMS. A minute quantity of TMS is added as an internal reference after the sample has been dissolved in a solvent free of interfering protons, often $CDCl_3$ (Nazarski, 2023). A small, cylindrical glass tube that is suspended between the magnetic poles serves as the sample cell. To make sure that the magnetic field is consistent throughout the solution, the sample is rotated on its axis. A coil connected to a 60-MHz radiofrequency (RF) generator is also located in the magnet gap. The electromagnetic energy required to alter the proton spin orientations is provided by this coil. A detection coil is positioned perpendicular to the RF oscillator coil. The detector coil does not receive any of the energy released by the RF oscillator coil when there is no energy absorption occurring. However, as the sample absorbs energy, the nuclear spin reorientation creates an RF signal in the detector coil plane, which the instrument records as a resonance signal, or peak. The various proton types in a molecule precess at different frequencies while the field strength is constant. The CW NMR spectrometer employs a constant-frequency RF signal and modifies the magnetic field intensity instead of altering the frequency of the RF oscillator to enable each proton in a molecule to reach resonance. All of the proton precessional frequencies rise with increasing magnetic field strength. Resonance occurs when a particular type of proton precessional frequency hits 60 MHz. The field strength can be increased by up to 20 parts per million (ppm) by the altering the current flowing through the pole pieces. All of the various proton types in the sample are brought into resonance by systematically altering the field in this manner. The magnetic field gets stronger as the pen moves from left to right. Each chemically unique proton type is represented as a peak on the chart as it enters resonance. The internal reference molecule TMS is responsible for the peak at $\delta = 0$ ppm. To get highly shielded protons to precess at 60 MHz, the field must be increased because they do so more slowly than unshielded protons. Therefore, on the chart, less shielded, or deshielded, protons appear on the left, and highly shielded protons appear on the right. It is common to refer to the area on the left of the chart as downfield, or at low field, and the area on the right as upfield, or at high field. Changing the radio frequency, or RF, is precisely the same as changing the magnetic field, as is done in a typical spectrometer. A 1-ppm change in the RF frequency has the same effect as a 1-ppm change in the magnetic field strength. Therefore, it is merely a question of instrumental design to alter the field intensity rather than the RF frequency.

1.9. CONTINUOUS-WAVE(CW)INSTRUMENTS

CW instruments are those that continuously change the magnetic field while scanning from the downfield end of the spectrum to the upfield end(Rohrer et al., 2001). This kind of spectrum is referred to as a frequency-domain spectrum as the chemical shifts of the peaks in it are computed using frequency differences from TMS. A CW spectrum can be identified by a unique feature.

After the instrument has scanned across a peak produced by a CW instrument, a decreasing series of oscillations known as ringing takes place. Ringing happens as a result of the excited nuclei not able to return to their equilibrium condition before the instrument field and pen have moved to a new location. The relaxation rate of the stimulated nuclei is slower than the scan rate. They continue to produce an oscillating, quickly fading signal as a result, which is captured as ringing. In a CW instrument, ringing is preferred since it is thought to be a sign that the field homogeneity is properly tuned. When a peak is a sharp singlet or a single, isolated peak or ringing is obvious.

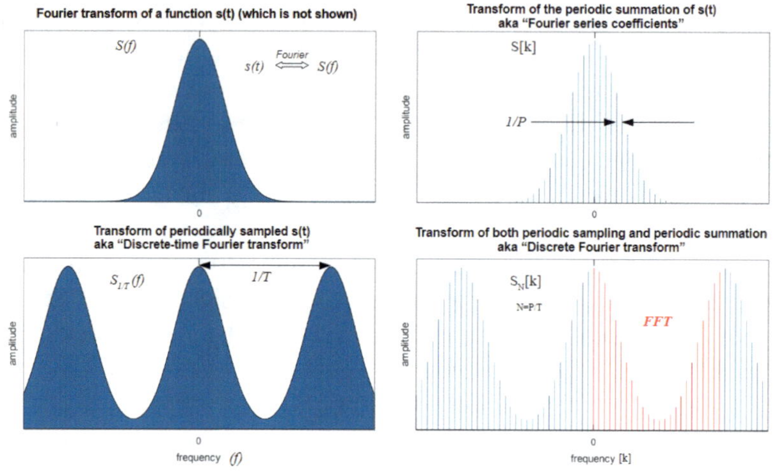

Figure 1.8. Fourier transform (FT) analysis.

Source:https://upload.wikimedia.org/wikipedia/commons/5/5a/Fourier_transform%2C_Fourier_series%2C_DTFT%2C_DFT.svg, Creative Commons License.

The CW type of NMR spectrometer works by excitation of the isotope nuclei at a time. Each unique proton type like phenyl, vinyl, methyl, and others, is activated separately in the case of $_1$H nuclei, and its resonance peak is recorded separately from the others. One type of hydrogen is examined first, followed by another, and so on, until all of the types have reached resonance. An intense but brief energy burst known as a pulse is used as an alternate method, which is typical of contemporary, high-tech devices, to activate all of the magnetic nuclei in the molecule at once. All of the $_1$H nuclei in an organic molecule are subjected to experience resonance simultaneously. This is achieved by a short-term (1–10 sec) burst of 90 MHz energy in an instrument with a 2.1 Tesla magnetic field. A pulse is produced by quickly turning the source on and off. A variant of the Heisenberg Uncertainty Principle states that if the pulse duration is extremely short-term, the oscillator frequency is uncertain because it was not long enough to establish a stable fundamental frequency, even though the oscillator frequency is set to 90 MHz. The pulse actually consists of a spectrum of frequencies centered on the fundamental frequency. All of the different kinds of hydrogens in the molecule can be excited simultaneously by this single energy burst due to the wide range of frequencies. When the pulse is stopped, the excited nuclei relax, or return to their initial spin state, as they start to lose their excitation energy. Every agitated nucleus releases electromagnetic radiation as it relaxes. The molecule emits a wide range of electromagnetic radiation frequencies at once because it has many distinct nuclei. This is called as free-induction decay (FID) signal. FID intensity decreases over time as each nucleus eventually loses its energy state. The FID can be rather complex and is a superimposed mixture of all the frequencies emitted. We use a computer and a mathematical technique known as a Fourier transform (FT) analysis (Figure 1.8)to extract the specific frequencies caused by various nuclei(Minhas and Oliver, 2022).

We may avoid the inherent complications of the Fourier transform and better grasp the process if we examine a very basic chemical, like acetone. The FID curve is made up of a single sinusoidal wave since acetone contains only one type of hydrogen. As the nuclei relax and their signal weakens over time, the signal decays exponentially. The FID is also referred to as a time-domain signal since time is the horizontal axis on this signal. The signal would look as a sine or cosine wave of constant strength if its intensity did not decrease. This measured wavelength, λ (difference between the maxima), can be used to determine its frequency. The frequency that the methyl hydrogens emit is not precisely the same as the frequency that was determined. The basic frequency of the pulse and the frequency of the acetone resonance differ because of the instrument design. Within a molecule, every proton present in a chemically identical environment is chemically equal and frequently displays the same chemical shift. Tetramethylsilane(TMS) and benzene, cyclopentane, and acetone are molecules with protons that are equivalent by symmetry considerations(Raut et al., 2021). All of the protons in these molecules have resonance at a single value. In its NMR spectra, (Figure 1.9)each of these compounds produces a single absorption peak. In the parts of the electromagnetic spectrum known as the ultraviolet (UV) and visible (VIS) areas, that is, the regions where wavelengths fall between 190 and 800 nm, the majority of organic molecules and functional groups are transparent. As a result, absorption spectroscopy is not very useful in this wavelength range. There are instances in which we can extract valuable data from these spectrum regions.

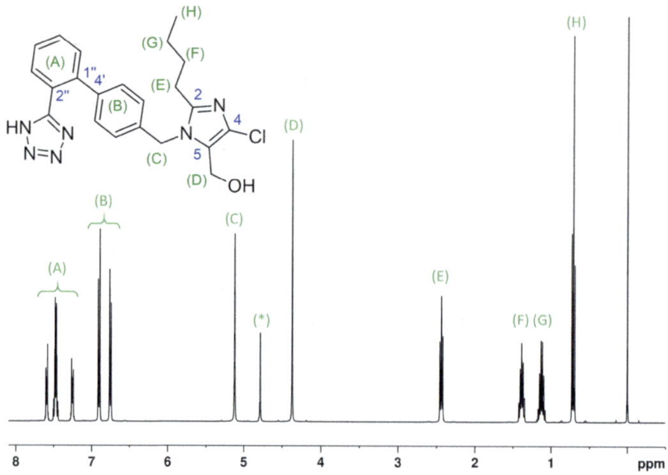

Figure 1.9. NMR spectra.

Source: Kuballa, T., Kaltenbach, K. H., Teipel, J., &Lachenmeier, D. W. (2023). Liquid nuclear magnetic resonance (NMR)spectroscopy in transition—from structure elucidation to multi-analysis method. *Separations, 10*(11), Article 11. https://doi.org/10.3390/separations10110572, Creative Commons License.

A fraction of the radiation may be absorbed as it travels through a transparent material continuously. If that happens, the remaining light produces an absorption spectrum, a spectrum with gaps in it, when it is passed through a prism. Atoms or molecules move from a low-energy state (ground state) to a higher-energy state (excited state) as a result of energy absorption. The energy difference between the excited and ground states is precisely equal to the energy of the absorbed electromagnetic radiation. When it comes to visible and ultraviolet spectroscopy, (Figure 1.10)the transitions that cause electromagnetic radiation in this spectrum to be absorbed are those between electronic energy levels.

An electron is moved from an occupied orbital to an unoccupied orbital with higher potential energy as a molecule absorbs energy. The transition or change from the highest occupied molecular orbital or HOMO to the lowest unoccupied molecular orbital or LUMO is probable. Most compounds have electronic level energy variations between 125 and 650 kJ/mole(Verma, 2023). The s orbitals, which correspond to s bonds, are the molecular orbitals with the lowest energy that are occupied by the molecules. The nonbondingor n orbitals, which have unshared pairs, stay at even higher energy levels than the p orbitals, which are at slightly higher levels.

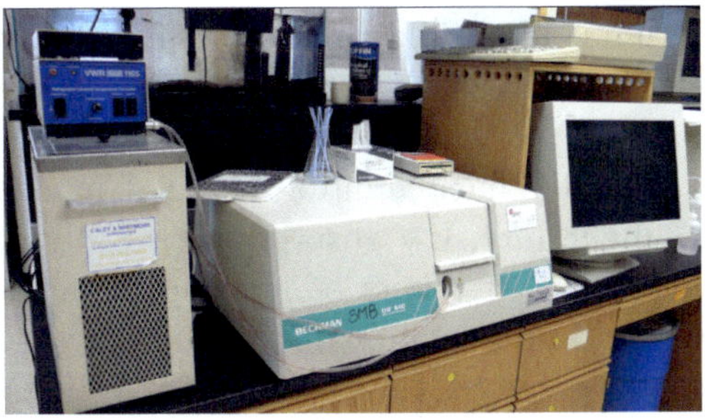

Figure 1.10. Visible and ultraviolet spectroscopy.

Source: https://upload.wikimedia.org/wikipedia/commons/e/ed/DU640_spectrophotometer.jpg, Creative Commons License.

The highest-energy orbitals are the empty, or antibonding, orbitals (π,σ). It is evident that a transition from the ground state's highest occupied energy level (HOMO) to its lowest unoccupied energy level (LUMO) requires less energy than a transition from a lower occupied energy level. The transitions that result in an electron's spin quantum number changing during the transition are prohibited; these are known as forbidden transitions.

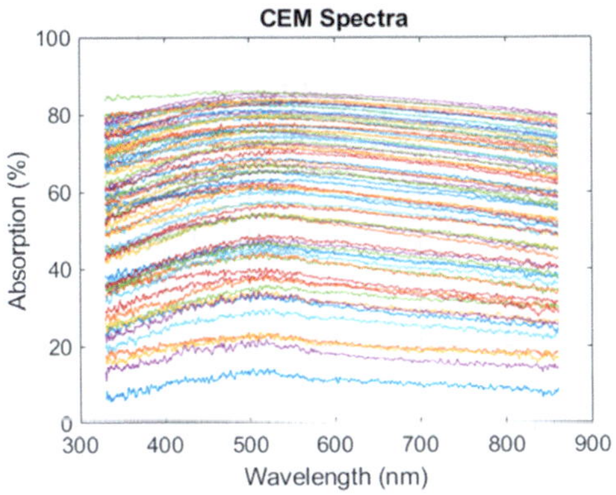

Figure 1.11. Absorption spectrum.

Source: Wacogne, B., Vaccari, N., Koubevi, C., Belinger-Podevin, M., Robert-Nicoud, M., Rouleau, A., &Frelet-Barrand, A. (2022). Absorption spectra description for T-cell concentrations determination and simultaneous measurements of species during co-cultures. *Sensors, 22*(23), Article 23. https://doi.org/10.3390/s22239223, Creative Commons License.

The absorption spectrum (Figure 1.11)of an atom that absorbs in the ultraviolet can occasionally have extremely sharp lines, which is to be expected for a quantized process that takes place between two distinct energy levels. However, because molecules typically have multiple excited modes of vibration and rotation at ambient temperature, UV absorption typically occurs throughout a wide range of wavelengths. Members of a group of molecules are typically in a variety of vibrational and rotational excitation states. These states have relatively closely spaced energy levels, which translate into energy disparities that are significantly less than those of electronic levels. The electronic levels are superimposed with the vibrational and rotational levels. A molecule may experience simultaneous vibrational, rotational and electrical excitation. Each electronic transition is made up of many lines that are so tightly spaced apart that the spectrophotometer is unable to resolve them because there are so many possible transitions, each of which differs from the others by only a small amount. The combined transitions show that a molecule's UV spectrum often consists of a wide band of absorption concentrated close to the major transition wavelength. The light absorption increases with the number of molecules that can absorb light at a certain wavelength. The extent of light absorption increases with a molecule's ability to absorb light of a certain wavelength. When a single species is responsible for the observed absorption, the Beer-Lambert Law is strictly followed. However, in certain situations, such as when the ground electronic state and a low excited state are in thermal equilibrium, when different forms of the absorbing molecule are in equilibrium, when solute and solvent form complexes through some kind of association, or when fluorescent compounds or compounds altered by irradiation are present, the law might not be followed.

Although photodiodes are also utilized in contemporary devices, a photomultiplier tube is typically used as the detector. The light coming from the light source is divided into two beams in a conventional double-beam instrument: the reference beam and the sample beam. The detected light is assumed to be equal to the light (intensity) entering the sample I_0 when there is no sample in the reference beam. The material used to build the sample cell needs to be transparent to the electromagnetic radiation employed in the experiment. Cells made of glass or plastic are typically appropriate for spectra in the visible spectrum. However, because they absorb ultraviolet light, glass and plastic cannot be utilized for measurements in the UV portion of the spectrum. Since quartz does not absorb radiation in this area, quartz cells must be utilized instead. To rotate the monochromator and provide a scan of all desired wavelengths, a mechanical system is needed. Because of the slow operation of this kind of equipment, recording a spectrum takes a long time. The diode-array spectrophotometer is a contemporary advancement over the conventional spectrophotometer. A diode array is made up of many photodiode detectors placed adjacent to one another on a silicon crystal. Every diode is made to capture a certain range of the spectrum. To record the full spectrum at once, the diodes are linked. This kind of detector records spectra rapidly and has no moving parts. A computer can receive its output, process it, and produce a number of helpful outputs. However, the benefits of this kind of instrument exceed the resolution loss for a large number of applications. The ability of a solvent to affect the wavelength of UV light that will be absorbed by stabilizing either the ground state or the excited state is arequirement for a good solvent. Polar solvents raise the energy of electronic transitions in polar molecules because they are less likely to form hydrogen bonds with the excited states of polar molecules than with their ground states.

1.10. TRANSITIONS

Transitions of the $n \rightarrow \pi^*$ type are shifted to shorter wavelengths by polar solvents(Mohammad-Jafarieh et al., 2021). The excited states may occasionally create stronger hydrogen bonds than the similar ground states. Since the energy of the electronic transition is reduced in such a situation, a polar solvent causes an absorption to shift to a longer wavelength. The $\pi \rightarrow \pi$ type transitions are shifted to longer wavelengths by polar solvents. The nuclei that the electrons hold together in bonds play a significant influence in deciding which wavelengths of radiation are absorbed, even though the excitation of electrons from ground to excited states causes the absorption of ultraviolet radiation. The energy gap between the ground and excited states is influenced by the nuclei, which also govern how strongly the electrons are bound. A chromophore is the collection of atoms that cause absorption. It is anticipated that the precise energy and strength of the absorption will alter in tandem with structural changes in a chromophore. It is frequently very difficult to predict how the absorption will vary as the chromophore structure changes. The only electronic transitions that can occur in molecules with only single bonds and no atoms with unshared electron pairs, such alkanes, are those of the $\sigma \rightarrow \sigma$ type. Because of their high energy, these transitions absorb ultraviolet light at wavelengths that are far shorter than those that can be measured analytically with standard spectrophotometers. Transitions of the $n \rightarrow \sigma$ type become significant in saturated compounds that contain atoms with nonbonding pairs of electrons. Although they are also rather high-energy transitions, they do absorb radiation that falls within the range that may be measured experimentally. Organic thiols and sulfides absorb between 200 and 220 nm, whereas alcohols and amines absorb between 175 and 200 nm(Monga and Nandini, 2024). The majority of absorptions are not visible in solution spectra because they fall below the cutoff values for common solvents. Unsaturated compounds with elements like nitrogen or oxygen may experience

$n \rightarrow \pi$ transitions. These transitions are among the extensively researched, especially when it comes to carbonyl compounds. These transitions are fairly susceptible to chromophore substitution. Around 280 to 290 nm, the typical carbonyl molecule experiences a $n \rightarrow \pi$ transition(Borjian et al., 2025). Since most $n \rightarrow \pi$ transitions are prohibited, their intensity is uncertain. The location and strength of a chromophore absorption band are altered when substituent groups are attached to a basic chromophore structure in place of hydrogen. Although the existence of the substituent groups alters the absorption of the primary chromophore, they may not directly cause the absorption of ultraviolet light.

1.11. AUXOCHROMES

Auxochromes are substitutes that raise the absorption strength and wavelength. Methyl, alkoxy, hydroxyl, halogen, and amino groups are examples of common auxochromes. There are different ways that additional substituents can affect absorption. A shift to a longer wavelength or lower energy is known as a bathochromic shift, (Figure 1.12)or red shift. A shift to a shorter wavelength or higher energy is known as a hypsochromic shift, or blue shift(Khan et al., 2022). An intensification, or hyperchromic effect is another method.

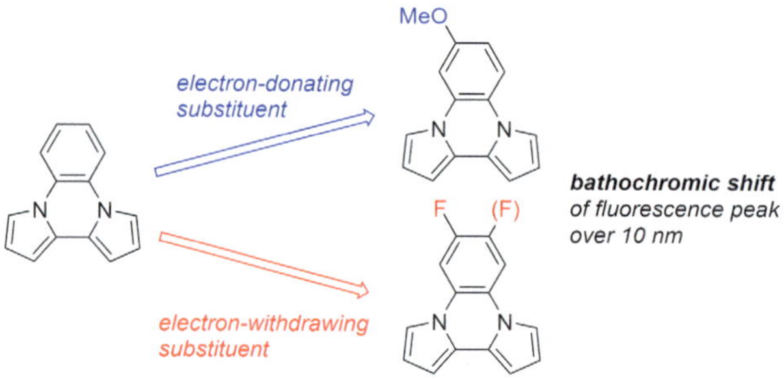

Figure 1.12. Bathochromic shift.

Source: Matsumoto, S., Takamori, M., &Akazome, M. (2023). Bathochromic shift of fluorescence peak in dipyrrolo[1,2-a:2′,1′-c]quinoxaline by introducing each of electron-donating and electron-withdrawing substituent. *Molecules, 28*(7), Article 7. https://doi.org/10.3390/molecules28072896, Creative Commons License.

A drop in intensity is known as the hypsochromic effect(Figure 1.13). Increasing the degree of conjugation in a double-bonded system is one of the best methods to induce a bathochromic shift. A chromophore electronic energy levels converge when conjugated double bonds are present. As a result, the wavelength of the light absorbed extends and less energy is needed to create a transition from an occupied electronic energy level to an unoccupied level. In addition to producing a bathochromic shift, the conjugation of two chromophores intensifies the absorption. Because conjugation moves the selective light absorption of isolated chromophores from a region of the spectrum that is not easily accessible to a region that can be easily studied with commercially available spectrophotometers, these two effects are crucial for the use and interpretation of electronic spectra of organic molecules. The degree of conjugation in the system can be connected with the precise location and strength of the conjugated system absorption band.

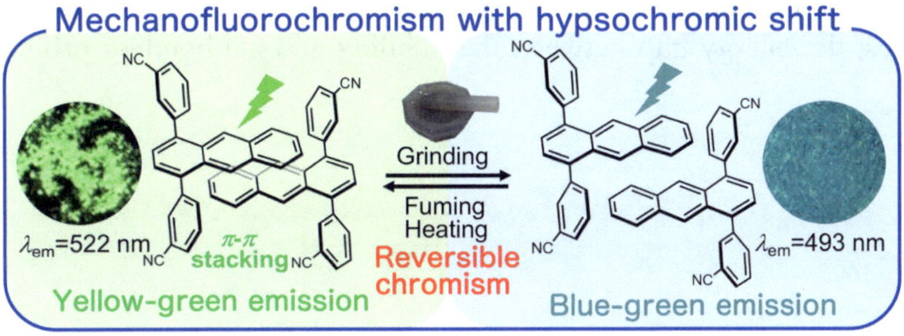

Figure 1.13. Hypsochromic shift.

Source: Kannen, F., Adachi, T., Nishimura, M., Yoza, K., &Kusukawa, T. (2024). Mechanofluorochromicproperties of 1,4-diphenylanthracene derivatives with hypsochromic shift. *Molecules, 29*(2), Article 2. https://doi.org/10.3390/molecules29020407, Creative Commons License.

It is implied that an increase in conjugation lowers the energy needed for electronic excitation by the bathochromic shift that arises from a conjugated system expansion. This is accurate, and molecular orbital theory provides the simplest explanation. The atomic p orbitals on each carbon atom combine to form p molecular orbitals, in accordance with molecular orbital theory (MO). We have two atomic p orbitals, f1 and f2, in the case of ethylene (ethene). By using linear combinations of these two p orbitals, we create two p molecular orbitals, (Ψ_1, Ψ_2)(Whangbo et al., 2021). The wave functions of the two p orbitals are added to produce the bonding orbital Ψ_1 and these two wave functions are subtracted to get the antibonding orbital Ψ_2. The energy of the antibonding orbital is higher than that of the original p orbitals, while the energy of the new bonding orbital, a molecular orbital, is lower.

The formation of two molecular orbitals can be observed as a result of the combination of two atomic orbitals. There were two electrons, one in each of the p orbitals of the atom. The new p system has two electrons as a result of the combination. These electrons wind up in Ψ_1, the bonding orbital, and form a new p bond since we fill the lower-energy orbitals first. In this system, the electronic transition is a $\pi \rightarrow \pi$ change from (Ψ_1, Ψ_2). 1,3-butadiene has four atomic p orbitals that make up its p system of two conjugated double bonds. The transition from the highest occupied molecular orbital (HOMO) to the lowest unoccupied molecular orbital (LUMO) has less energy as the number of p orbitals that make up the conjugated system increases. As conjugation increases, the energy gap between the bonding and antibonding orbitals gets smaller.

CHAPTER 2

RAMAN SPECTROSCOPY

2.1. INTRODUCTION

Since light is an electromagnetic wave, it interacts with matter through the charge around atoms or molecules since it is made up of oscillating electric and magnetic fields. The interaction between light and matter is studied as an electron-photon interaction, but it can be seen as a radiation-matter interaction. Understanding how electromagnetic radiation (EMR) couples and exchanges energy with the electrons that make up the substance it interacts with through its electric field is essential to understand Raman scattering (Brittain, 2006) (Figure 2.1).

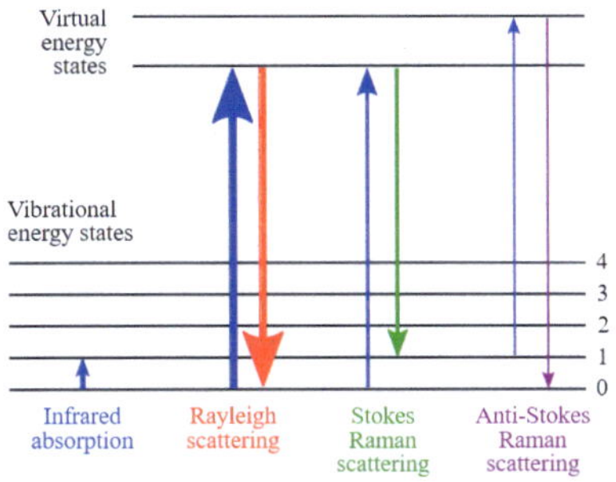

Figure 2.1. Raman scattering.

Source: https://upload.wikimedia.org/wikipedia/commons/4/41/Raman_energy_levels.svg, Creative Commons License.

Many physical phenomena and technologies, including lasers, photovoltaic cells, and different types of spectroscopy, including Raman spectroscopy, depend on our understanding of photons. For resolution and power, other devices like filters and detectors are used. The kind of laser that is employed is determined by how well it balances fluorescence interference and scattering efficiency and is appropriate for a variety of samples, including biological materials and inorganic chemicals, whether they are liquid or solid. Shorter wavelengths require a larger excitation energy, which can produce more fluorescence but produces stronger Raman signals (Wei et al., 2015). This is because the scattering power is inversely proportional to the wavelength raised to the fourth power. Depending on the sample, longer wavelengths are recommended to reduce the impacts of fluorescence. These wavelengths have higher penetration into the sample and less heating, making them perfect for biological samples and other materials that are prone to fluorescence. However, because of their lower Raman scattering efficiency, they usually need more sensitive detection equipment. It is crucial to know that shorter wavelengths cause more fluorescence and stronger signals, but because of their high photon energy, which increases scattering power and may produce more heat, they can also degrade the sample. For biological samples, chemical molecules, and specific minerals, certain wavelengths might be needed. Selecting the right laser wavelength is essential to maximizing Raman spectroscopy results for various sample types. Raman spectroscopy uses the following standard laser wavelengths: 473 nm (blue), 532 nm (green), 633 nm (red), 785 nm (near infrared), and 1064 nm (near infrared) (Philippidis et al., 2021).

2.2. INTERACTIONS

Depending on the area of interaction, phenomena that occur when light hits a substance can be categorized as either surface or volume phenomena. Because of the discontinuity caused by the medium it interacts with, the electromagnetic wave experiences reflection and transmission effects through its fields in contact with the surface. The fraction of photons reflected and transmitted, as well as phenomena pertaining to interfacial fields and surface reconstructions, were investigated using this discontinuity, which is connected to various refractive indices. Information regarding the arrangement of atoms, molecules and symmetry can be collected from the fraction of light that enters the sample volume. The wavelength of the laser light, the power output, the pulse duration or the time over which the power is measured are some of the variables that affect how many photons are in a laser pulse or beam. Electrons that are bound to crystal atoms and do not participate in electrical conduction are known as bound electrons. This interaction can be seen in the visible, ultraviolet, and a portion of the near-infrared spectrums. Electrons that contribute to electrical conduction are known as free electrons or holes, and this interaction are visible in the mid- and far-infrared(Harbecke et al., 1996). Because of their ionic charge, atoms and ions interact with electromagnetic radiation. This interaction is known as lattice scattering, and it is visible in the mid-, far-, and microwave regions(Sun et al., 2016). Although the electrons in dielectric materials are bonded to the atoms, an electric field has the ability to displace them. In an oscillator model, bound electrons oscillate around an equilibrium point when disturbed by an external force, such as an applied electric field that corresponds to the incident electromagnetic radiation. The friction is related to velocity and force is proportional to displacement. The dielectric function and polarizability in quantum physics are determined similarly to how they are in the classical Lorentz model(Figure 2.2).

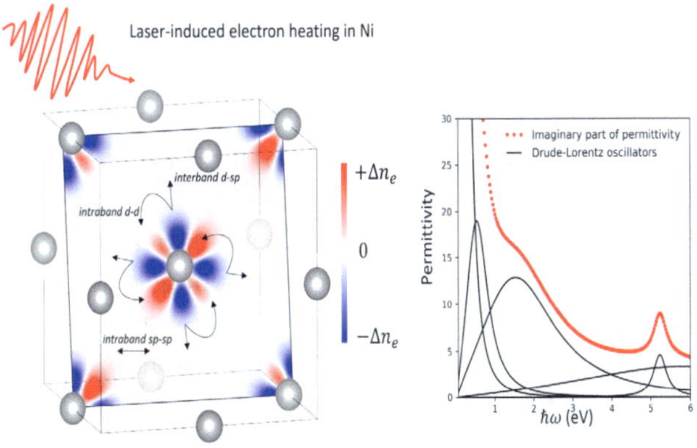

Figure 2.2. Lorentz model.

Source: Silaeva, E., Saddier, L., &Colombier, J.-P. (2021). Drude-lorentz model for optical properties of photoexcited transition metals under electron-phonon nonequilibrium. *Applied Sciences, 11*(21), Article 21. https://doi.org/10.3390/app11219902, Creative Commons License.

Assuming the ground state, we can examine the interaction of a free atom with an electromagnetic field. As a result, we only need to take into account the effects of polarization and absorption, not emission. A model used in optics to explain the frequency dependence of a refractive index is called the Sellmeier oscillator(Moghaddam and O'Leary, 2020).

The refractive index and wavelength are related by the Sellmeier equation. This is essential for developing optical systems, (lenses, prisms), and is especially helpful for describing optical materials across a range of wavelengths. The wavelength of the incident light has a significant impact on the penetration depth. Shorter penetration depths are usually the result of faster absorption of shorter wavelengths (higher energy photons). The strength with which a semiconductor absorbs light at a certain wavelength is measured by the absorption coefficient α, a property for materials.

2.3. OPTICAL DENSITY (OD)

The OD parameter can also be used to quantify an optical medium absorption coefficient (α)(Myers et al., 2013). The design of optical filters depends on this parameter. The physical measure of a material absorption per unit length is called its OD. The transparent optical materials are employed for this purpose. The interaction of radiation with matter, or how materials behave when exposed to light, is highly dependent on the incident wavelength (λ). The amount of light that is absorbed by a sample as it travels through it is measured by absorbance (A), also referred to as OD. In spectrophotometry, it is a central idea. An optical density filter (ODF), sometimes referred to as a neutral density (ND) filter, is a kind of filter that lowers light intensity without altering its color(Vargas, 2025a). The ND filter function is to decrease the amount of light that enters the lens, giving the user more control. This is helpful in a number of situations, including long-exposure photography, preventing overexposure, and decreasing depth of field in bright lighting. ND filters come in various varieties according to their optical density. The optical density of fixed ND filters is fixed. With variable ND filters, the filter elements can be rotated to change the optical density. By permitting longer exposure durations, ND filters are used to create motion blur in moving objects like clouds, water, etc. With larger apertures, they are also utilized to regulate depth of field under bright lighting. The ND filters are employed in a variety of optical investigations to regulate light intensity precisely without changing the light characteristics. Glass or resin covered with a neutral substance that uniformly attenuates light across the visible spectrum is commonly used to create ND filters(Pfaender, 1996). Superior ND filters guarantee that the image color balance does not shift. A dimensionless indicator of light attenuation over a path length is OD. The absorption coefficient, which has units of inverse length, measures the attenuation per unit length. OD is determined by dividing the incident light intensity by the transmitted light intensity over a given path length. The attenuation of light per unit length of the material is described by the absorption coefficient. For ceramic materials and the majority of semiconductors, laser energies of about 50 mW are often utilized(Sanghera et al., 2012). However, certain oxides, such as manganese oxides and low melting point polymers, require the use of laser energies of about 5 mW. Neutral density filters with an OD of about 1.0 are required in these situations. Preliminary measurements of the incident laser powers on the sample are crucial. Using powers of about 50 mW and concentrating on a representative peak in the response spectra of the sample is a smart strategy. Usually, this peak is called the control peak. It is possible to determine whether the curve changes with each repeat of the scan by capturing the spectrum surrounding the control peak. In order to improve the signal-to-noise ratio, the power should be decreased if this is the case until the curve stays constant. Rayleigh scattering processes are crucial when using lasers in spectroscopic procedures.

2.4. SCATTERING

Shorter wavelengths (blue light) scatter more than longer wavelengths (red light) because the strength of Rayleigh scattering (Figure 2.3)is inversely related to the wavelength (fourth power)(Lockwood, 2023). The scattered photons have the same energy (wavelength) as the incident photons since this scattering is an elastic process. Short wavelengths cause strong elastic scattering in inhomogeneous materials and the majority of binary oxides, which amplifies fluorescence and protects the Raman signal.

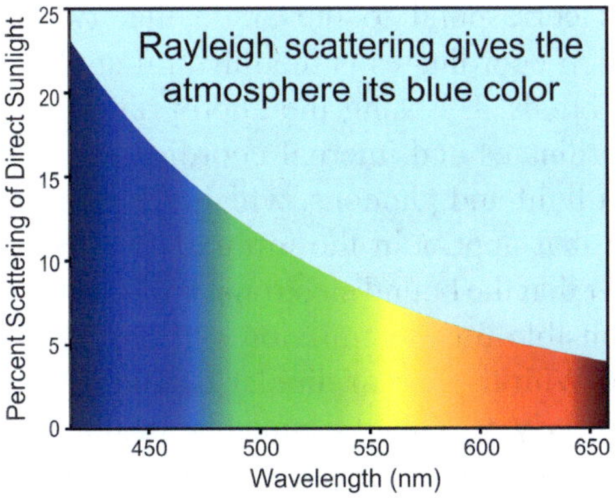

Figure 2.3. Rayleigh scattering.

Source: https://upload.wikimedia.org/wikipedia/commons/d/dd/Rayleigh_sunlight_scattering.svg, Creative Commons License.

From a quantum perspective, electromagnetic radiation is seen as a collection of particles known as photons, but from a classical perspective, it is seen as coupled perpendicular electromagnetic fields. It is crucial to utilize the equation that connects photon energy to wavelength in optical spectroscopy. A substance is said to be crystalline if its atoms are arranged in regular patterns that repeat on a regular basis, creating a perfect lattice. We can describe harmonic crystals when these atoms vibrate around equilibrium points, which are described by periodic oscillations and expressed by sine or cosine functions known as harmonic functions. From a mathematical perspective, generalized coordinates are simple; however, the functional expression of the potential energy must be known in order to calculate the Lagrangian(Sadeghi, 2023). The goal of the experiment, however, is to comprehend the bond strength, spatial geometry, and the environment in which the molecule is immersed. The oscillation frequencies of the molecules as well as other data, such as line width and behavior as a function of temperature, are known. Although they enable a sophisticated and rigorous mathematical approach, generalized coordinates do not clearly depict what occurs in the molecule.

It only produces partial results because displacement coordinates are established based on the direction of the bonds. A full set of coordinates, known as internal coordinates, that are appropriately positioned within the molecule are introduced via another formalism. Internal coordinates fall into one of the following categories: Stretching coordinates that correspond to the measured elongation or extension of the link between the individual atoms. The angle created between the bonds is related to the bending coordinates. Out-of-plane bending coordinates correspond to the angle that two distinct bonds form with the plane. Torsion coordinates is the dihedral angle variation determined by three successive bonds. By using the appropriate transformation, the link between normal coordinates and internal coordinates can be established. The relationship between light and phonons, which are atom-to-atom vibrations in the crystal structure that appear in the infrared frequency range, in solids is examined. Remember that the bound electrons or those creating the bonds inside the crystal are responsible for the emission and absorption in the visible and ultraviolet spectrum. With the use of dipolar oscillators, the primary behavior of phonons may be clearly explained from a classical perspective. It enables us to comprehend how strongly polar substances absorb infrared light.

Scattering, or the displacement of light from its initial direction of incidence on the substance, is a physical process that occurs when electromagnetic radiation interacts with matter(Björn, 2015). A functional expression of the emission frequency as a function of the wave vector describes this scattering mechanism. Oscillating electric dipoles are created when the incident electric field vector interacts with the electrons, causing the waves to scatter as a result of the internal electron oscillations. With two primary forms of scattering, the oscillating electric dipoles create new emission sources that re-emit radiation in all directions. Rayleigh scattering is the term for the radiation that emerges from a material when the frequency of the light emitted equals the frequency of the incident light. When a material emits light with a frequency that differs from the incident light frequency, this is known as inelastic scattering. We can perceive objects because Rayleigh scattering is the most likely. The fourth power of the incident wavelength has an inverse relationship with scattering efficiency.

2.5. RAMAN SCATTERING

Raman scattering is the inelastically scattered portion of light. Stokes Raman scattering occurs when the energy of the scattered light is less than that of the light incident on the material; anti-Stokes Raman scattering occurs when the energy of the scattered light is greater than that of the incident light on the material. Forced oscillators that depict the dispersion of electromagnetic radiation with atomic patterns will show that the model is appropriate and provides a good depiction(Stenholm, 1973). The interaction of incident light with the electrons within the material explains Raman scattering. Since the energy of the light is frequently insufficient to excite these electrons to a higher electronic energy level, the result is a change in the molecular vibrational or rotational state. We can refer to the inelastic light scattering process, which contains the Raman process, as part of the radiation-matter interaction. Themolecules interact with the photons when light strikes it, triggering some of their inherent vibrational modes and causing the molecule to oscillate in that particular mode. There are multiple intrinsic or inherent normal modes of vibration in molecules that make up the material. When Indian physicist Chandrasekhara Venkata Raman was investigating how light scatters in alcohols in 1923, they noticed that the color of the light traveling through these solutions changed(Mukherji and Mukhopadhyay, 2018).

They then filtered the scattered radiation in order to separate the Rayleigh component. It took until 1928, five years later, to satisfactorily explain this new radiation. In 1930, he received the Nobel Prize in Physics for this scientific contribution. The scattering process that occurs when electromagnetic radiation, particularly in the visible spectrum, interacts with materials includes: Rayleigh scattering, which is intense and has the same frequency as the incident radiation. This indicates that the process is known as an elastic process since the energy of the incident photon and the energy of the output photon are equal. The extremely weak Raman scattering is roughly 10^{-6} in relation to the incident radiation intensity(Gel'mukhanov and Ågren, 1999). The fact that the incident and radiated photons have different energy is one crucial feature. A photon is used in the inelastic scattering process of Raman spectroscopy. It involves the interaction of two particles: a molecule's electron and a photon. Normal modes like rotation, vibration, or a vibro-rotational combination are activated by this interaction. Energy conservation is maintained throughout this process. A new photon with energy E_f is created when a molecule interacts with an incident photon with energy E_i through the electrons that make up the molecule. Similar to the absorption spectroscopy methods, there are other Raman spectroscopy methods that make use of this inelastic scattering. Two photons are involved in Raman scattering: one incident photon that is absorbed and another photon that is released from the material. Both Stokes Raman and anti-Stokes Raman scattering are feasible in this two-photon process that occurs in a single phase(Romeike et al., 2015). The photon that is released has a lower energy in the Stokes process and a higher energy in the anti-Stokes process. Experimental behavior can be closely visualized using classical theories. Sometimes the approximation is sufficient, and other times it is not as accurate but offers an interpretation from a terminological perspective. In order to accomplish this, the Raman effect can be interpreted in the context of a diatomic molecule, emphasizing key properties such electrical polarization. Radiation is seen as an electromagnetic wave in the visible spectrum, which is represented by laser radiation, in the radiation-matter interaction.

The macroscopic polarization of the medium as a result of all the dipole moments the electric field induces in the material replaces the dipole moment μ in complex systems like solids, particularly crystals. An inelastic scattering known as Raman scattering occurs when the induced dipole moment oscillates at a frequency distinct from the incident radiation. Because some of the photon energy has been transmitted to the interacting medium, this phenomenon is known as Stokes scattering if the photons scatter at frequencies lower than the excitation field. Because the material medium releases energy when photons scatter at higher frequencies, this phenomenon is known as anti-Stokes scattering. Raman scattering can reveal details about the sample composition and structure. It can be deduced that this quantum of energy has been utilized for the transition between two energy levels existent in the sample by applying the energy conservation principle and taking into account inelastic interactions, in which energy is either transferred or absorbed by the sample. Each material has a distinct frequency spectrum because of its crystalline structure, bonding, impurities, etc., as determined by measuring the energy difference between incident and scattered photons, which is equal to the energy difference between the affected vibrational levels. Each Raman spectrum serves as a fingerprint that uniquely identifies the sample being examined. The excitation light has no effect on the Raman frequencies that the material emits. About 0.1 percent of the light incident on the sample in scattering phenomena is an elastic process, which happens more frequently than an inelastic one, in which only one in 10^6 or 10^7 incident photons is Raman(Kuzmany, 2009). Since 99% of the molecules in room temperature Raman scattering are in the lowest energy vibrational state, it is more likely that the sample will absorb energy to activate one of the normal vibrational modes and emit a lower frequency photon. As a result, Stokes scattering is roughly 100 times greater than anti-Stokes scattering. If the internal fields (of molecule that is excited) are larger than the electric field of the radiation that excites it, the linear approximation is correct. The capacity of the excitedelectron cloud to be distorted or deformed by the external electric field is measured by α. The C = C and C = O systems provide a example of the importance of polarizability in the Raman signal intensity. A strong Raman signal is linked to this normal mode of vibration in the C = C system when the vibration happens and the bonds stretch, producing a strong polarization as a result of the electron clouds being displaced. The vibration results in a lower Raman signal for the C = O system because it causes a lesser displacement of the electron cloud.

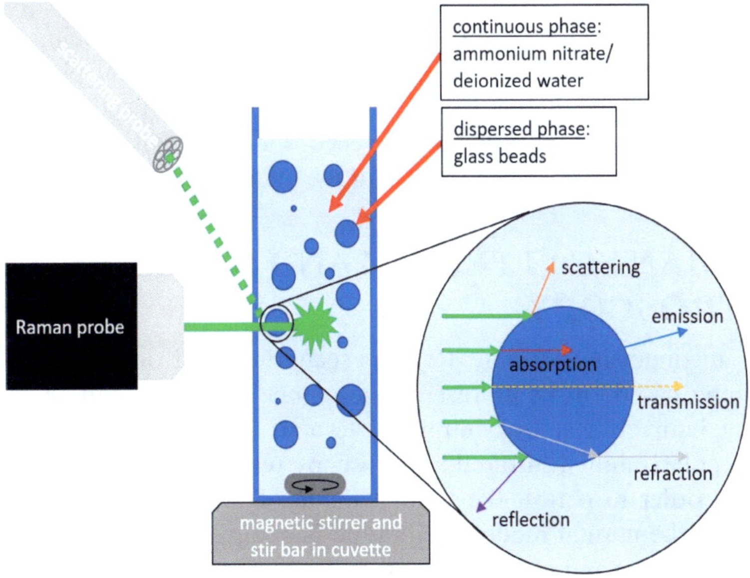

Figure 2.4. Raman spectroscopy.

Source: Spoor, E., Oerke, V., Rädle, M., &Repke, J.-U. (2024). Raman spectroscopy of disperse systems with varying particle sizes and correction of signal losses. *Sensors, 24*(10), Article 10. https://doi.org/10.3390/s24103132, Creative Commons License.

This rule has a practical application in the study of polymers, where Raman spectroscopy (Figure 2.4)is used to describe the carbon chains that make up the polymer backbone and infrared spectroscopy is used to collect information about the functional groups of the polymer chain. Only diatomic or triatomic molecules can be accurately determined if a certain mode of vibration is Raman-active based on the fluctuation of polarizability during molecular vibration.

Group theory must be applied, taking advantage of any potential symmetries in the molecule, to ascertain whether a vibration in molecules with more intricate atomic arrangements is Raman-active(Laane and Ocola, 2012). A weak Raman band is produced by a vibration that has no visible effect on polarizability. Therefore, very weak Raman scattering is typically produced by strong polar compounds like the O-H bond. Bonds with electron clouds dispersed over a wide area, like double or triple bonds in carbon chains, are the cause of strong Raman scattering. An external electric field can readily modify the electron clouds of these bonds. Such a bond electron density is significantly altered when bent or stretched, which causes the induced dipole moment to fluctuate as well. While the intensity of light scattering, including Raman scattering, depends on the wavelength of the incident light, with the scattering behavior being inversely proportional to λ^4, the energy difference between the incident photon on the sample and the scattered photon corresponds to the energy needed to activate a normal mode of vibration, and this difference is independent of the wavelength of the incident light used as the source for the material analysis.

2.6. RAMAN SCATTERING AND IR ABSORPTION SPECTROSCOPY

We can distinguish between Raman scattering and infrared absorption spectroscopy based on the aforesaid factors, one of which is the possible for generating both effects. In contrast to Raman scattering, which requires a change in polarizability, infrared spectroscopy requires a change in the dipole moment in order to permit the normal mode of vibration. The condition of activation of the normal mode of vibration is compatible with the differences between the Raman and infrared spectra in terms of the modes that can be seen as well as the signal intensity. The fact that the Raman and infrared modes are mutually exclusive for molecules with a center of inversion is another crucial feature. It should be noted that infrared spectroscopy typically measures the strength of the absorbed signal, while Raman spectroscopy captures a dispersed signal at various angles with respect to the incident light source. The analyte of interest, which is connected to the selection rules, determines the technique to be used for a given analysis. The symmetrical energetic positions of the anti-Stokes and Stokes lines with respect to the Rayleigh line suggest that the energy for the transition is the same in both situations. The Stokes line, however, is substantially more intense than the anti-Stokes line at room temperature. When an incident photon causes a molecular transition from a lower state to a higher state, leaving the molecule in an excited state, the Stokes line is produced. Since the transition in the anti-Stokes line takes place from a higher state to a lower state, the molecule must be excited and then relaxed to a lower state, typically the ground state. Since there are more molecules in the ground state than in the excited state at room temperature, there is a chance that the molecule will go through a Stokes-type transition. The Boltzmann distribution quantitatively determines the relative population of molecules between the ground state and the excited state(Lasorne et al., 2011). The cross section is a crucial parameter in the analytical handling of signals since it strongly depends on the kind of material being examined. The likelihood that an incident photon will be dispersed as a Raman photon is proportional to the Raman cross section σ_j. The molar absorptivity derived from Beer's law, which represents the cross section for optical absorption, is one of the frequently used quantities in optical spectroscopy. The integral cross section of the Raman band under study needs to be calculated in all directions and across all wavelength ranges since the cross section σ_j includes scattering in all directions.

This suggests that in order to evaluate PR, light must be collected around the sample at a solid angle of 4π steradians (sr) and integrated over the whole Raman band(Qi et al., 2025). Only a limited range of solid angles and one or more directions of light scattering by the sample are seen because of the way the measurement device is configured. Therefore, defining a differential Raman cross section, $\beta = d\sigma_j/d\Omega$, where Ω is the solid angle utilized to capture the signal, is more feasible(Gu et al., 2000). β is measured in $cm^2mol^{-1}sr^{-1}$. Comparing the Raman signal of a known or standardized sample quantitatively to an unknown cross section that acts as a normalization parameter is the useful method to know the Raman cross section. The spectrometer response function, where the ratio of the peak areas of the two bands under examination equals the ratio of the cross sections, is used to calculate β. Other signals, such as optical absorption and fluorescence, are processed in Raman intensity detection and compete to lower the signal-to-noise ratio. Many technical features and experimental methods concentrate on lowering the fluorescence signal because the cross section of these signals is 6–8 orders of magnitude larger than the Raman cross section.

The theory of quantum mechanics explains processes at the microscopic level for various states of matter. By adding average parameters, as in the case of crystal field theory, the behavior of atomic motifs, such as solitary molecules, may be examined in the solid state, and the entire solid can be taken into consideration. The isolated behavior of molecules is a good initial approximation to infer their behavior within the solid. It will demonstrate how molecules produce the Raman scattering effect. A virtual state and two photons, the incident photon and the scattered photon, are involved in Raman scattering. Second-order time-dependent perturbation theory must be used in order to establish the quantum explanation. Detectors for Raman spectroscopy are essential for picking up (recording) the dispersed signal, which has a signal-to-noise ratio of roughly 1 in 10^7. In comparison to other optical signals, this one is weaker. The interaction between the molecules in the materials being studied and the input laser light produces the weak scattered signal that the detector records. When the incident radiation interacts with the regular vibrational modes, a scattered signal known as Raman scattering is produced. This dispersed light is transformed by the detectors into a detectable voltage or current signal, which is further filtered by optical systems and amplified and adapted electronically by the electronic equipment.

The Raman spectrum, whose strength is related to the number of interactions that take place between the incident photons and the molecules, can be analyzed due to the signal that has been detected. The near-infrared range (NIR) spans from approximately 12,500 cm^{-1} to 4,000 cm^{-1}. This region is appropriate for the studying organic compounds because it primarily contains the C–H, N–H, and O–H bands(Weyer, 1985). Due to the vibrations of the H atoms in the corresponding molecules, such as C–H, S–H, N–H, and O–H, many applications in organic compounds in the NIR region typically involve functional groups. Two subregions, one between 4,000 and 1,300 cm^{-1} and another between 1,300 and 650 cm^{-1}, comprise the mid-infrared region, which spans 4,000 cm^{-1} to 650 cm^{-1}. The primary absorption bands in the first area are dependent on the functional group they originate from rather than the entire molecule structure.

The movement of the bonds holding the groups to the rest of the molecule is related to the second region. These are the frequencies at which single bonds in polyatomic molecules deform and vibrate when bent. The 700–300 cm^{-1} range is known as the far-infrared region (FI). Molecular vibrations in solids and liquids, lattice vibrations and molecular vibrations in crystals, and pure rotations and vibration-rotation transitions in gaseous molecules are the causes of this region. The molecular structure and content of both organic and inorganic materials can be inferred using Raman spectroscopy. It is a flexible and, in theory, nondestructive method that makes it simple to examine both solid and liquid materials.

Gases can also be analyzed, but only in certain conditions depending on the density and sample container or holder. The minimum sample size is milligrams, and even less. One instrument can be used to obtain Raman spectra, which are only vibrational spectra. Compared to infrared techniques, sample preparation is typically much easier and typically does not require sample preparation previously. Non-resonant Raman spectroscopy in the visible and near-infrared is one of the many variations of Raman spectroscopy that are currently available.

2.7. FOURIER TRANSFORM RAMAN SPECTROSCOPY (FTIR)

FTRS, (Figure 2.5)one of the Raman spectroscopy variations intended to increase the signal-to-noise ratio, also reduces sample fluorescence, enhances frequency precision, and permits low-frequency detection(García-González et al., 2013). The radiation intensity is measured as a function of wavenumber in conventional Raman spectroscopy.

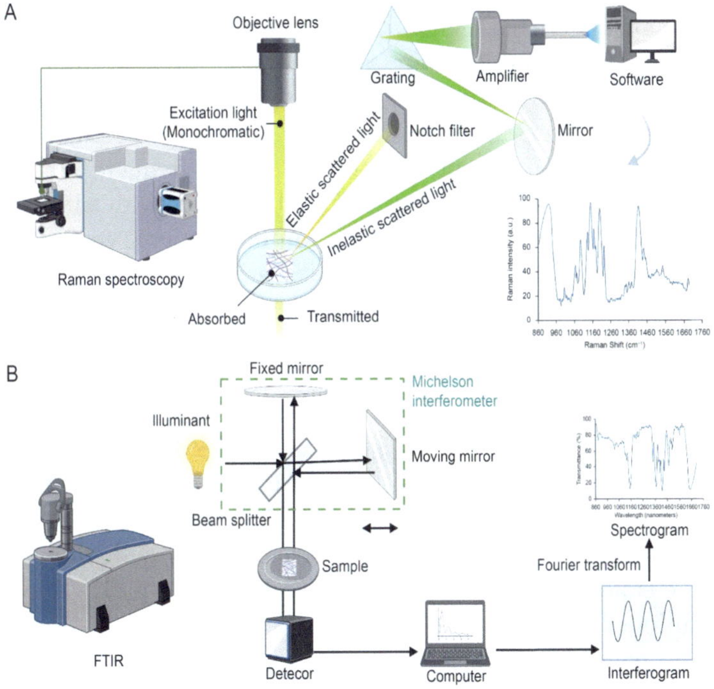

Figure 2.5. Fourier transform Raman spectroscopy (FTRS)(A and B).

Source: Lu, L., Zhao, Y., Li, M., Wang, X., Zhu, J., Liao, L., & Wang, J. (2024). Contemporary strategies and approaches for characterizing composition and enhancing biofilm penetration targeting bacterial extracellular polymeric substances. *Journal of Pharmaceutical Analysis, 14*(4), 100906. https://doi.org/10.1016/j.jpha.2023.11.013, Creative Commons License.

Light intensity at multiple wavelengths is simultaneously recorded in instruments that use FT. CCD camera-type photodetectors are used in this process; they use the Fourier transform to transform the spectrum into a conventional one. There are some differences between FTIR spectroscopy and Raman spectroscopy, two vibrational methods for studying molecular structure.

The inelastic scattering of light (in which the incident photon energy is different from the scattered photon), is the principle in Raman spectroscopy. It is referred to as a Stokes-type interaction if the scattered photon energy is low and an anti-Stokes-type interaction if it is high. The emitted photon has less energy than the incident photon because some of the energy from the incident photon is used to activate the normal vibrational modes of the molecules that make up the sample when coherent radiation, typically from a laser source, strikes it.

There is roughly one Raman photon for every 10^7 laser photons in the scattered Raman light. This technique needs high-sensitivity detectors, noise reduction, electronic amplification, and suitable optical filtering. Raman spectroscopy quantifies the amount of light that the sample scatters. FTIR spectroscopy is an optical absorption method in which molecules in a sample absorb IR radiation. The molecules absorb particular frequencies linked to the allowed typical vibration modes when infrared radiation with a broad-spectrum range strikes the sample. A broad-spectrum infrared source as well as infrared sensor-based filtering and detecting devices are needed for this technique. Depending on the thickness of the sample to be examined, the technique can be used with either signal transmission or reflection. In Raman spectroscopy, vibrational modes are triggered when a molecule's polarizability changes, such as when an external electric field deforms the electron cloud. Vibrational modes in FTIR spectroscopy are triggered when the dipole moment changes, separating the positive and negative charges inside the chemical structure. Raman spectroscopy makes it possible to analyze samples without any prior preparation, which makes it possible to research gases, liquids, and solids. Since water has no effect on the Raman signal, many liquids are observed in aqueous solutions.

Prior sample preparation is usually necessary for FTIR spectroscopy (potassium bromide). Water absorption in the infrared area can produce significant interference with the spectra, which is one of the key issues because it is an absorption process. Symmetrical molecules or bonds with a constant dipole moment (C=C or S–S bonds) are ideal for Raman spectroscopy(Fleming and Williams, 2019). Polar molecules or bonds with a significant change in dipole moment (O–H or C=O bonds) are the ideal candidates for FTIR spectroscopy. Because water is so polar, this makes analyzing aqueous samples more difficult. Although some sample preparation techniques, such as crushing the material into KBr pellets, may change the sample, FTIR spectroscopy is likewise non-destructive. Raman spectroscopy is used to investigate molecular structure, crystal formations, and chemical composition in the fields of chemistry, biology, and pharmaceuticals. It is helpful in areas including nanotechnology and forensic analysis. A variation called surface-enhanced Raman spectroscopy (SERS) uses metal substrates or nanoparticles to amplify the Raman signal(Cialla et al., 2012). To identify functional groups, investigate chemical bonds, and perform routine quality control, FTIR spectroscopy is widely employed in organic and inorganic chemistry, pharmaceuticals, polymer science, and biochemistry. FTIR and Raman are complementary methods. FTIR is better at detecting polar molecules and functional groups in a variety of samples, whereas Raman is better suited for nonpolar compounds and samples in aqueous conditions. The RAMAN system uses a high-energy monochromatic light source to illuminate the sample under study. Recent developments in optoelectronic device technology have led to the usage of lasers with varying wavelengths. Photons are scattered after interacting with atomic motifs, such as molecules or chemical complexes, on the sample. Elastic and inelastic collisions make up the dispersion. Rayleigh scattering is the term for the elastic scattering of photons that takes place in all directions while preserving the photon initial energy.

When the polarizability of the atomic motif changes, a form of inelastic scattering known as Raman scattering takes place in which the incident photon energy is different from the scattered photon. A powerful monochromatic radiation beam passing through a sample of molecules that are susceptible to a change in polarizability causes the Raman effect. The electron cloud has to be readily distorted during the vibrational moment. The dipole moment of the molecule must permanently alter as a result of vibrations in the infrared spectrum. A lower energy photon is released when a molecule in Raman scattering accepts the energy of the incident radiation and is excited to a higher state (Stokes). One of the regular vibration modes of the molecule must be activated by the energy differential between the incident and scattered photons. Rayleigh and Raman scattering processes have an efficiency of 10^3 and 10^7 when conducted in the right experimental settings(Jones et al., 2019). Only one photon of the Raman type is inelastically scattered out of a million photons that incident on the sample; the otherphotons are elastically dispersed. High-intensity excitation sources are necessary because of the process inefficiency. Energy is transferred to the vibrational modes of the molecules in the sample via the oscillating induced dipole. The polarized molecule stops oscillating as the electromagnetic wave interacts with it, and in this quasi-state, it returns to its ground level, emitting energy in all directions. The photon or scattered radiation leaves the sample with less energy than the incident one because a quantum of radiation is still present in the molecule; the energy difference between the incident and scattered photons is used to calculate the quantum required to initiate a normal vibration mode. A vibrational quantum of energy may also be transferred during the interaction process with the incident photon if the scattering molecule is already in an excited vibrational state. In the event of a change in polarizability, the molecule may emit a photon with a higher energy than the incident photon, putting it in a lower vibrational state. We refer to this process as anti-Stokes. Since there are more molecules in the ground state than in the excited state at room temperature, the Stokes process outperforms the anti-Stokes process. The Raman shift in both procedures is proportional to the transitional normal vibration mode. The fingerprint is represented by the spectrum corresponding to the examination of a sample, which is made up of a series of lines energetically situated at the value corresponding to one of the typical vibration modes. A laser that excites the sample and a spectrometer, which usually uses diffraction gratings with 600, 1,200, 1,800, and 2,400 lines per millimeter, make up Raman spectroscopy(Rzhevskii, 2024). The laser radiation enters the spectrometer through the rear, which is specifically made for this purpose. In certain areas of the spectrometer, it travels in the same direction as the sample Raman radiation. An entry slit of a double monochromator or, occasionally, a triple monochromator is the target of the Raman emission. A stepper motor demarcates the radiation based on its wavelength, and the motor directs the wavelength, which in certain situations may be a photomultiplier cooled by the thermoelectric effect to improve the signal-to-noise ratio. CCD cameras are frequently utilized to detect the Raman signal. In theory, non-destructive and micrometric-sized particles can be examined using the Raman Microscopy version.

Figure 2.6. Confocal Raman microscopy.

Source: https://upload.wikimedia.org/wikipedia/commons/9/98/Confocal_Raman_imaging_microscope_Witec_alpha300_.jpg, Creative Commons License.

FTIR and Raman spectroscopy offer a powerful tool for characterizing materials. FTIR is best suited for examining polymeric functional groups, whereas Raman is best for inorganic species and low-frequency studies. For some samples, the fluorescence issue isavoided by using the He-Ne laser with a wavelength of 632 nm. The depth of laser radiation penetration can be used for both qualitative and quantitative analysis when varying excitation wavelengths are employed because the penetration depth is roughly proportional to the excitation light wavelength. This makes it possible to examine multilayer structures.

2.8. CONFOCAL RAMAN MICROSCOPY

Confocal Raman Microscopy,(Figure 2.6) however, can be used to do similar investigations. It is possible to employ lasers from the visible (632, 532, and 473 nm), ultraviolet (244 nm), and near-infrared (785–1,064 nm) regions(Giridhar et al., 2017). Since only one of the about 10^7 incident photons is a Raman signal, the laser strength is crucial. The Raman signal may be obscured by the fluorescence background, which can be increased by raising the laser power, depending on the sample properties. The system configuration, including the laser utilized, influences the region of the Raman spectrum where higher resolution is required, which influences the choice of Raman signal detection method. The resolution, bandwidth, and reaction time detectors used in Raman spectroscopy vary. The cost of high-performance detectors, which use sophisticated technology, varies according to the needs of the operator and the analysis needed. CCDs, or charge-coupled devices have high sensitivity, good quantum efficiency, and capacity to pick up weak signals, CCDs are the widely used detectors in Raman spectroscopy and are frequently seen in industrial systems. They can be combined with electronic amplification and counting systems for appropriate electronic filtering, and they are used in high-resolution approaches. CCDs are perfect for dispersive Raman spectrometers because they can detect avariety of wavelengths at once. They are either internally cooled or need external cooling. PMTs, or photomultiplier tubes, are utilized in certain applications. PMTs can detect extremely weak Raman signals because of their high sensitivity to low light levels.

They havebeen preferred due to their reaction to light intensity. In scanning systems, PMTs are frequently employed in conjunction with monochromators and provide quick response times. Electron-Multiplying CCDs (EMCCDs) has increased sensitivity. EMCCDs are highly sensitive to low light conditions because they amplify the signal before reading it. EMCCDs are especially helpful in applications that need for the detection of weak Raman signals, including in low-light or low-concentration materials. Near-infrared (NIR) Raman spectroscopy is performed using Indium Gallium Arsenide (InGaAs) detectors(Barbillat and da Silva, 1997). Benefits include their sensitivity to longer wavelengths, whereas conventional CCDs may hardly to detect them. Avalanche Photodiodes (APDs) are employed in single-photon detection and other applications requiring extremely high sensitivity. APDs are frequently employed in Raman spectroscopy time-correlated single-photon counting (TCSPC) systems. Back-Illuminated CCDs has increased quantum efficiency. By positioning the photodetector on the back of the chip, these CCDs lower absorption and reflection losses and boost quantum efficiency. Better sensitivity is provided by back-illuminated CCDs, particularly in the visible and ultraviolet spectrums. The spectrum range,sensitivity, and particular application or user requirements are some of the elements that influence the detector selection. In order to eliminate the laser signal, which functions as disturbing incident radiation, more filters are required. Since a monochromator is one of the basic components of spectroscopic techniques, it is critical to comprehend its primary purposes. In optical spectroscopy techniques, the monochromator is used to choose particular scattered light wavelengths. The intensity of the radiation scattered after interacting with the molecule being studied is isolated in Raman spectroscopy.

2.9. MONOCHROMATOR

Only Raman (inelastically scattered) light is permitted to flow through the monochromator, which filters out Rayleigh scattered light (elastic scattering), which has the same wavelength as the incident light. This Raman light, which is aimed at the detector and contains information about the inelastic interaction, has a variable wavelength because of molecular vibrations. The monochromator reduces optical noise and undesired signals by choosing particular wavelengths, which enhances spectral resolution. Depending on the application and desired spectral range, different types of monochromators may be employed. Monochromators ultraviolet Raman spectroscopy is comprised of shorter wavelengths, whereas those for near-infrared spectroscopy are adjusted for longer wavelengths. The monochromators in Raman spectroscopy are those with diffraction grating. The number of lines per mm or millimeter determines the resolution of a grating, which is a set of lines produced using holographic techniques. Reflective gratings are the used, while transmissive gratings are also an option. The grating separates the light spectrum spatially, much like a prism would. These monochromators split light into its individual wavelengths using a diffraction grating. The desired wavelength can be directed toward a collimator and subsequently to the detector by rotating the grating.

Prism monochromators (Figure 2.7)are less popular and distribute light using a prism in accordance with wavelength-dependent refractive index. They perform well and are frequently employed in settings where a broad spectral range is not as important. Double monochromators are employed when extremely high resolution and improved stray light rejection are necessary. It is crucial to remember that the intensity of the radiation that is collected drops, so it might be necessary to boost the incident laser signal intensity on the sample. Double monochromators, which combine two monochromators enhance signal purity and reject stray light even more effectively.

Figure 2.7. Prism monochromators.

Source:https://upload.wikimedia.org/wikipedia/commons/d/dc/Prism_monochromator%2Cwith_light_path_illustrated%2Cfrom_the_laboratory_of_AHNU%2CMar_2017.jpg, Creative Commons License.

They are especially helpful for applications that need very low noise levels, like analyzing very small samples, finding traces of a sample in a matrix, or detecting very weak Raman signals. A strong spectroscopic method called Surface-Enhanced Raman Scattering (SERS) increases the Raman scattering of molecules adsorbed on rough metal surfaces or nanostructures(Tian et al., 2002). Inelastic scattering results from the interaction of light with molecular vibrations, which causes Raman scattering. A molecular fingerprint is produced by the energy difference, which correlates to the vibrational modes of the molecule. Localized surface plasmon resonances (LSPRs) are caused when light strikes a rough metal surface or metal nanoparticles(Mayer and Hafner, 2011). The metal conduction electrons oscillate coherently, producing strong electromagnetic fields at the surface. The Raman scattering of molecules close to the tissue surface is enhanced by the electromagnetic fields. There will be a transfer of charge between the metal surface and the adsorbed molecule. The electrical characteristics of the molecule may be altered by the contact, increasing the cross sections for Raman scattering. Silver (Ag), copper (Cu), and gold (Au) are often utilized metals for SERS substrates. The high plasmonic characteristics of these metals in the visible and near-infrared portions of the electromagnetic spectrum are the reason they were selected. Colloidal solutions of metal nanoparticles, such as silver or gold nanoparticles, are called nanoparticles. Metal nanoparticles combined with other materials (silicon or graphene) improve stability and functionality are known as nanocomposites. SERS is useful in environmental monitoring, food safety, and medical diagnostics since it can detect small amounts of chemical and biological substances. Single-molecule detection is made possible by SERS sensitivity, which offers information on single-molecule molecular dynamics and interactions.

2.10. SERS

Adsorption, catalysis, and corrosion processes are among the surface phenomena that are studied using Material Science SERS. Forensic Science SERS helps identify and detect traces of forensic materials, including inks (stains), explosives, and narcotics. Both liquid and solid organic and inorganic materials are included in the samples that are measured. While the majority of samples do not require conditioning before analysis, some do require. One of the key variables in the measurements is the laser power, which is controlled by a set of neutral density filters tailored to the particular laser being used. The laser utilized in this instance had a wavelength of 473 nm, and the system consisted of a confocal microscope for accurate laser impact area localization and sample alignment, a current-cooled CCD detection camera, and a 1m monochromator. For target selection, the system provides 1 μm lateral resolution in the range of 10× to 100×. The number of repetitions, wavelength range, and sampling duration can all be chosen using the Raman system. An important factor is sampling duration; longer sampling periods can broaden spectral peaks because they heat the sample from laser exposure while simultaneously increasing signal intensity. Inorganic oxide samples may occasionally deteriorate as a result of this heating during the measuring process. Since monocrystalline silicon has a peak at about 520.7 cm^{-1}, it is typically used to calibrate Raman devices(Kadlečíková et al., 2022). Due to its flexibility in producing doped semiconductors for use in rectifier diodes, emitter diodes, and a variety of other devices, including transistors, silicon is one of the commonly used materials in device technology. The solar cells, which are essential for contemporary energy generation, is another important use for silicon-based junctions. The second most prevalent substance in the crust of the Earth is silicon. Silicon has been the subject research in Raman spectroscopy due to its distinct Raman peak and significance in the semiconductor industry. The silicon Raman spectrum shows a visible peak that offers important details on the electrical and structural characteristics. The material used to calibrate our Raman system is silicon. At 520.7 cm^{-1}, silicon displays a sharp Raman peak that corresponds to the optical phonon mode in its crystal structure. Raman spectroscopy is used in the semiconductor industry to track the quality of silicon wafers as they are being manufactured. Stress or impurities in the silicon crystal can be detected by the location, width, and strength of the Raman peak. For example, changes in the 520.7 cm^{-1} peak can be used to infer stress-strain analysis; compressive deformation usually results in a shift to higher wavenumbers, whereas tensile deformation leads the peak to shift to lower wavenumbers(Vargas, 2025b). The way semiconductors behave in relation to temperature is another crucial feature. This characteristic is helpful in the non-contact thermal analysis of silicon-based devices since the Raman peak typically broadens and changes toward lower wavenumbers as the temperature rises. Raman spectroscopy is used in the microelectronics sector to observe thin films, find residual stress, and guarantee silicon integrity throughout processing at different phases of device manufacture.

Figure 2.8. Lasers.

Source: https://upload.wikimedia.org/wikipedia/commons/2/23/Lasers.JPG, Creative Commons License.

2.11. LASERS

Lasers (Figure 2.8)with wavelengths that optimize Raman scattering efficiency and decrease fluorescence are frequently utilized for silicon research. Since silicon samples typically do not need to be specially prepared for Raman analysis, the method is practical and non-destructive. But the surface needs to be perfect and clear of impurities like dust and grease. One of the most significant semiconducting materials in terms of technology is zinc oxide. Its bandgap is one of its important characteristics for optoelectronic applications; it usually ranges around 3.14 eV and is highly dependent on the growth method. Four atoms make up each unit cell of ZnO, a semiconductor material with a hexagonal wurtzite structure. It is currently unclear how to interpret the usual vibrational mode, which is situated at about 580 cm^{-1}. It has been documented because of a local vibrational mode (LVM) of nitrogen and an E_1(LO) mode associated with oxygen vacancies(Fujita et al., 2007). The precursor source (NH_4OH), which is added as an impurity into the ZnO crystalline structure at high treatment temperatures when the hydroxides attached in the initial growth stage transform into ZnO, is responsible for the nitrogen presence in the samples, which becomes visible at temperatures above 150°C. The usual vibrational modes associated with an ZnO crystalline structure may or may not be apparent due to the development of impurities or symmetry breakdown, depending on the synthesis techniques and whether the sample is in powder or film form. Vanadium pentoxide (V_2O_5) and its hydrated forms (V_2O_5 nH_2O) crystallize in a laminar structure and are used in a variety of technological applications, including electrochromic devices, solid-state battery cathodes, electronic and optical switches, and catalysts in chemical reactions. Amorphous films of vanadium oxide undergo physical changes when exposed to methanol, ethanol, acetone, and isopropanol. The Raman spectrum displays distinctive bands that represent the typical vibration modes of Zn and Cr atoms in tetrahedral and octahedral environments made of oxygen atoms, which are situated approximately 400 and 900 cm^{-1}, respectively. Because of the interaction between the Zn and Cr atoms, the bands about 941 cm^{-1} are linked to vacancies in tetrahedral and octahedral sites and could be crucial for surface gas absorption. Due to the high temperatures used during the sintering process, these oxygen vacancies are produced.A semiconductor (Gallium arsenide) (GaAs) is used in electronics, especially in light-emitting devices and some sensors. Since it can expose details about the crystal's structure, quality, and the existence of defects or impurities, Raman spectroscopy is a useful technique for investigating the vibrational characteristics of GaAs. Understanding the molecular behavior of water depends on the vibrational modes. Depending on the phase (gas, liquid, or solid) and environmental factors like temperature, pressure, contaminants, or whether the water is diluted with another liquid, certain Raman shifts may differ slightly.

CHAPTER 3

FLUORESCENCE SPECTROSCOPY

3.1. INTRODUCTION

The emission of light by certain molecules occurs after the absorption of light. The electromagnetic nature of light, the molecular composition of the emitting molecule, and the surroundings of the luminescent molecule influence the absorption and emission processes. The absorption and emission of light by these molecules are closely linked to the chemical bonds that are formed between the valence electrons and orbitals that make up partially filled outer shells. Because of their repulsive qualities or properties, the electrons that occupy the filled inner shells of the atoms that make up these molecules are localized upon the atoms and contribute secondarily to the electronic structure of molecules. The electronic spectra of molecules in the visible and ultraviolet portions of the spectrum are caused by the orbitals and electrons in molecules whose energies are defined by the strong bonding interactions between atoms, including those valence electrons in atoms that do not have covalent bonding. The combination of occupied atomic orbitals is the source of a chemical bond or molecular orbital. The Pauli exclusion principle, i.e., a maximum of two electrons can occupy one orbital, governs the filling pattern of the molecular orbital, while the spatial direction of the overlap defines the type of chemical bonding (Krapp et al., 2006). A σ bond is created when two atomic orbitals that connect the bound atomic nuclei merge along the axis. In a σ bond, the two atoms (bonded) have a highly localized charge distribution. The two electrons that occupy the σ orbital can originate from either one atom or both from the same atom, even though each atom involved in the σ bond contributes one atomic orbital to the development of the σ orbital. The σ bond is referred to as a coordinate covalent bond in the latter scenario and a covalent bond in the former. The electron pair that makes up the σ bond will stay for more time close to one atom than the other if the two atoms that are linked by the covalent bond have uneven attractions. The bond is referred to as a polar covalent bond as the atom that is more firmly attracted to it is more electronegative than the other. Electronic repulsion and the exclusion principle prevent more bonds from forming between any two atoms in a molecule because the electronic charge in a chemical bond is localized between two atoms. The molecule normally binds electrons involved in σ bonding firmly. It takes a lot of energy to remove a single electron from the molecule or move it to a higher energy molecular orbital that is unoccupied. A 7t bond is created when two atomic orbitals overlap above and below the location where the bound atoms' nuclei are located. Since π bonding is a weaker connection than σ bonding, it occurs after σ bonding has reached saturation. Atomic p or d orbitals are always involved in the production of π bonds, while s orbitals are never. To reduce interelectronic repulsion, the electronic charge distribution in a π bond is localized outside the σ bond axis (Sobczyk et al., 2005). The π electrons are free to go throughout the molecule, outside the region of their atoms of origin, and may be distributed over multiple atoms since they are not as concentrated between the parent atoms as the σ electrons are.

3.2. BONDS

Instead of an alternating localized two-atom π bonds, a set of π orbitals that are dispersed throughout the atoms is generated when several atoms are σ bonded in succession and each atom possesses a p or d orbital with the correct spatial orientation to form a π bond with the others. π delocalisation occurs throughout the entire molecule in certain cyclic chemical compounds, especially organic compounds. These are the molecules most likely to exhibit fluorescence and are referred to as aromatic compounds. The ionization potential and electronic spectra of π electrons occur at lower energy than those of σ electrons because they are weakly bonded. The transition involving π electrons is in the vacuum ultraviolet (100 to 200 nm) or at the edge of the near ultraviolet for molecules with isolated π bonds(Hochstrasser, 2002).The π electron spectra of molecules with two or three conjugated delocalized π bonds are typically in the near ultraviolet (200 and 380 nm), whereas the π electron spectra of super-delocalized π systems, aromatic molecules, and widely delocalised linear molecules like carotenes range from the near ultraviolet for small molecules to the near infrared (200 and 1500 nm) for large molecules. There are electrons in the valence shell that are already paired in atoms with more than four electrons in the valence shell (nitrogen, oxygen, and Sulphur).Although the isolated atom valence shell is incomplete, the orbitals that these electrons occupy are already filled.Despite having energies similar to those of other electrons in the same shell, the electrons cannot form covalent bonds.They are referred to as nonbonding or *'n'* electrons.The valence shell electrons involved in the bonding process will have energies lower than the electrons themselves when the atom is bonding.Although they are not as low as the σ electrons, the energies of the electrons participating in π bonding will be lower than those of the n electron. Since the n electron has a higher energy than the σ or π electron, it must be taken into consideration as a possible contributor to the spectrum characteristics of molecules that include them. Non-bonded electron pairs in molecules can be moved into open molecular orbitals in acceptor molecules, even though they do not normally form covalent bonds.These compounds usually contain metal ions or atoms with empty valence shells.This leads to the creation of a coordination complex, where a coordinate covalent bond is created between the acceptor and the nonbonded electron pair donor.Although most coordinate covalent interactions are of the σ type, some chemical species can also form π -type coordination complexes, mostly the complexes of aromatic ligands with main group transition metals and their ions.One kind of coordinate covalent bond formation is Bronsted basicity.A hydrogen ion conjugate acid is formed when the Bronsted base adds two non-bonded electrons to its unoccupied 1s orbital(Shao et al., 2015).The nonbonded pair on the base loses identity as a result of the σ bond that forms between the hydrogen ion and the base.When interpreting the spectra of compounds with nonbonded electron pairs, the development of coordination complexes is important.

3.3. ORBITALS

Each molecule has a number of higher energy molecular orbitals that are normally empty. These orbitals are in addition to the low-energy molecular orbitals that contribute the chemical bonds of molecules. These orbitals are known as antibonding molecular orbitals and represent areas where electrons may be excited if enough energy is applied to the molecule. Each bonding orbital in a molecule has one antibonding according to molecular orbital theory. Antibonding orbitals can have either π symmetry, where the electron density is out of the plane that runs parallel to the bonding axis, and are referred to as π^* orbitals, or σ symmetry, where the electron density is along the bonding axis. The promotion of an electron from an initially occupied bonding or nonbonded orbital (σ, π, or n) to an initially vacant molecular orbital through energy absorption is known as electronic absorption(Wu and Zhu, 2013). The demotion of an electron from a σ^* or π^* orbital to a σ, π, or n orbital causes light emission. With the exception of free radicals, all organic compounds have bonding and nonbonding orbitals are doubly occupied in the ground electronic state, which is the state with the lowest electronic energy. The electrons in these orbitals have opposite spin quantum numbers.However, the initially occupied orbital and the antibonding orbital both become singly occupied when an electron is promoted to a higher orbital.Then, there is at least a chance that one electron will alter its spin, and the Pauli exclusion principle no longer requires that the two electrons have opposite spins. Each of the electrons that fill molecular orbitals has unique physical characteristics, such as spin angular momentum, energy, and orbital angular momentum. The orbital symmetry determines the electron's path within the molecule, which in turn determines the orbital angular momentum.Electrostatic attractions from the atomic nuclei in the molecule and repulsions from other molecular electrons influence an electron's energy in a molecule. The orbital that the electron occupies determines these attractions and repulsions. Nuclear attractions are stronger for electrons having σ bonds built from sp2 atomic hybrid orbitals than for electrons occupying orbitals built from sp3 hybrids.This is because s orbitals have a higher electron density near the nucleus than p orbitals, and the sp2 hybrid has high s character.The electron spinning on its own axis is the appropriate way to imagine the origin of the spin angular momentum. The electrical configuration of a molecule is the arrangement of its electrons among its molecular orbital. The resultant orbital angular momentum, spin angular momentum, and energy of each electronic configuration of a molecule are made up of the scalar sum of the energies of all of the electrons as well as the vector sums of the orbital and spin angular momenta, respectively. The electronic state of the molecule is described by its electronic energy and angular momenta.Only the π and n electrons in most organic molecules participate in spectroscopic phenomena in easily accessible regions of the electromagnetic spectrum. The repulsions of all other electrons, such as the σ electrons and the inner shell electrons localized on individual atoms, are typically ignored in order to simplify the treatment of molecular electronic structure.

3.4. TRANSITIONS

Spectroscopic transitions can only provide information on the energy differences between states, not the absolute energies of electronic states.A molecule's ground electronic state, which is the state with the configuration where all of its electrons are in the lowest energy orbitals available, is sometimes referred to as its lowest electronic energy state. one electrically stimulated state of the molecule is created when one electron is promoted from an orbital that is occupied in the ground state to one that is typically empty in the ground state. Each molecule can have a large number of electronically excited states, in part due to the large number of vacant orbitals.A specific distribution of electronic charge within a molecule characterizes each of its electronic states. This implies that a molecule's dipole moment in an electrically stimulated state will frequently deviate from its ground state dipole moment. A molecule's electronic states are categorized based on the orbital angular momenta and spin associated with each state. Electron spectroscopy (Figure 3.1)uses a number of classifications, but the Schonflies point-group and the Platt free-electron model are popular(Jellison and Joshi, 2018).

Figure 3.1. Electron spectroscopy.

Source: Taurino, A., &Carlino, E. (2023). The relevance of building an appropriate environment around an atomic resolution transmission electron microscope as prerequisite for reliable quantitative experiments: it should be obvious, but it is a subtle never-ending story! *Materials, 16*(3), Article 3. https://doi.org/10.3390/ma16031123, Creative Commons License.

Rather than focusing on different electronic configurations of the excited states, both of these methods attribute the primary differences in the energies of different absorption bands in an electronic spectrum to variations in the orbital angular momenta of the excited states of the transitions that give rise to the absorption bands. The rigorous group-theoretic base of the point-group system and the free-electron molecular orbital method of Platt's system are the areas of difference between the two approaches, which are only applicable to highly symmetrical homoatomic molecules. When analyzing the vectorial coupling of substituent electronic behavior with thefeatures of the electronic transitions intrinsic to the symmetrical parent molecule, absorption spectra then become an issue. In the case of unsubstituted aromatic hydrocarbons and polyenes, both approaches adequately explain observed spectra; however, they do not perform well when more than two strongly interacting substituents are joined to the parent molecule and less well when such molecules have weakly interacting substituents. The spin angular momentum is used to identify electronic states. The ground state of the majority of organic compounds contains an even number of electrons paired in the lowest molecular orbitals.An electron's spin angular momentum vector can be orientated in one of two ways, and the number of electrons with either orientation is equal in the ground states of the majority of organic compounds.As a result, there is no net spin angular momentum.

Singlet states are those with zero spin angular momentum, or no unpaired electrons, because the net spin magnetic moment of zero does not couple with the magnetic field when an external magnetic field is applied across the molecule.The magnetic field has no effect on the spectral lines or bands that result from the transition between these states; that is, their energy remains single or unsplit.

3.5. SPIN

An electron will typically keep its original spin if it is moved from an occupied orbital to a higher, previously vacant orbital. The excited state will therefore likewise be a singlet state. However, the molecule will contain two unpaired electrons if the promoted electron changes spin.When an external magnetic field is applied to a molecule containing two unpaired electrons, the resulting spin angular momentum vector (spin magnetic moment) can be orientated in three different ways: normal to the field vector, opposite to the field vector, and in the same direction as the field vector.Three states that are obtained from the identical electrical arrangement and orbital angular momentum will differ slightly in their magnetic energy.

A simple spectral line or band will divide into three when it transitions to or from these states.A molecule is therefore considered to be in a triplet excited state if it has two unpaired electrons.The multiplicity of the state, which is generally equal to the number of unpaired electrons (+1), is the number of orientations that the spin angular momentum vector can have in an external magnetic field. The triplet state that results from a particular electronic configuration in isolated molecules typically have lower energy than the singlet state of the similar configuration. This is because, in accordance with the Pauli exclusion principle, the promoted electron and the residual electron in the triplet state are lowers the triplet state repulsive energy.

3.6. ENERGY STATES

An excited state may be created in molecules with nonbonded electron pairs when a n electron is promoted to an empty π orbital.Singlet (n, π *) states and triplet (n, π *) states of this type are both possible. These excited states are known as n, π * states(El-Sayed, 2002).The n electrons are somewhat less energetic than atomic valence shell electrons due to the stabilization provided by the bonding interaction, even though the bonding π orbitals have a significantly low energy than the atomic orbitals from which they are created.Because of this, the energy gap between the n and π * (empty π) orbitals is smaller than the gap between the π and π * orbitals, and the n electrons in the same molecule have a greater energy than the π electrons.As a result, π , π * excited states involving the same π * orbital typically lie higher than n, π * excited states. Nitrogen heteroatoms in the aromatic ring and carbonyl or thiocarbonyl groups are the most common sites for orbitals in aromatic compounds.The n orbital, which is located on the heteroatom that carries the unshared electron pair with which it is related, is actually a sp2 hybrid that is directed in the aromatic ring plane or at right angles to the π and π * orbitals. The π and π * orbitals are not always perpendicular to unshared electron pairs.Some, like those found in hydroxyl and exocyclic amino groups, are nearly parallel to the aromatic ring π orbitals; in these situations, they might even be involved in the aromatic system's π -electron structure.

In general, the excited states created by the promotion of these electrons behave more like π , π * states than like n, π * states in some ways.Because they result from the transfer of an electron from the exocyclic group to the aromatic ring, these n, π *-like states shall be referred to as intramolecular charge-transfer states or l, π * states (Grabowski et al., 2003).The promoted electron may be moved far from its point of origin in the l, π * state, but in the n, π * state, it is primarily localised close to the group from where it originated. The l, π * state is a specific example of coordination complexes caused by metal-to-ligand electron transfer. The promotion of a σ electron from the aromatic ring to an empty σ * orbital located on an exocyclic group can also result in intramolecular charge-transfer excited states.When the carbonyl group is conjugated with the aromatic ring in aromatic carboxylic acids, aldehydes, and ketones, as well as in some transition metal complexes where σ electrons from the aromatic ligand can be transported into unoccupied d orbitals on the metal ion, this phenomena is frequently observed. Excited states can also arise from the promotion of σ -bond electrons to π * orbitals or from the promotion of σ , π , or n electrons to σ * orbitals.The spectroscopic transitions involving these orbitals typically occur in the vacuum ultraviolet region (200 nm), which is outside the range of absorption and fluorescence spectrophotometers. This is because the cr orbitals lie so low in energy and the σ * orbitals lie so high in energy.A molecule that absorbs light is in one of several vibrational levels of one of its excited states.The stages that return the excited molecule to its ground state is much slower, taking anything from 10^{-14} to several seconds, even though the absorption process is quick, takes about ten to fifteen seconds.One of the slowest processes in excited states is the return to the ground state via molecular luminescence, which takes 10^{-10} to a few seconds.Thermal equilibrium through vibrational energy loss, on the other hand, occurs more quickly, taking 10^{-14} to 10^{-12} seconds(Fultz, 2010). The electronic absorption spectra do not reflect vibrational processes in the excited state since they take place after light absorption.Similarly, vibrational excitation activities in the ground state are not reflected in the fluorescence spectra since they take place after light emission.

3.7. ABSORPTION SPECTRA AND FLUORESCENCE

The absorption spectra show the vibrational structure of the Franck-Condon excited state because all absorptions are thought to originate from the same vibrational level of the ground state and end in different vibrational levels of the electronically excited state.

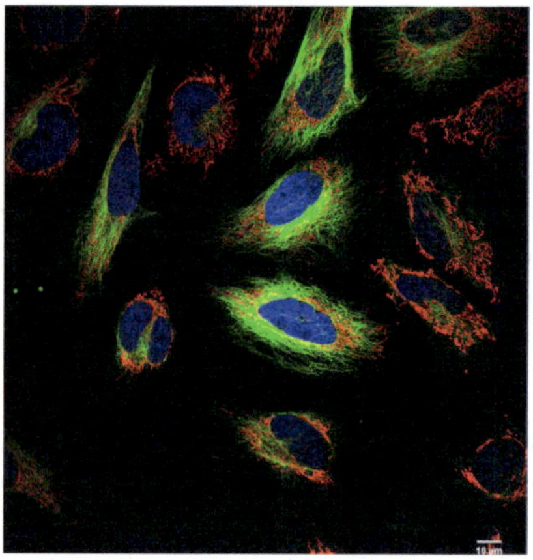

Figure 3.2. Fluorescence.

Source: https://upload.wikimedia.org/wikipedia/commons/5/57/Multicolor_fluorescence_image_of_living_HeLa_cells.jpg, Creative Commons License.

The fluorescence spectrum represents the vibrational structure of the Franck-Condon ground state since all fluorescence (Figure 3.2)originates from the same vibrational level of the lowest excited singlet state and ends in the different vibrational levels of the ground electronic state. The transitions between the ground state lowest vibrational level and the excited state lowest vibrational level should be represented by vibronic transitions in absorption and fluorescence spectra that have the same energy and match the longest wavelength vibronic feature of the wider or longest wavelength absorption band and the shortest wavelength vibronic feature of the fluorescence band.The fluorescence and longest wavelength absorption bands will be mirror images of one another if the vibrational structures of the Franck-Condon ground and lowest excited singlet states are identical. The fluorescence spectrum will be on the long wavelength side of the 0–0 band, while the absorption spectrum will be on the short wavelength side.However, because the ground and excited states have different charge distributions and equilibrium geometries, their vibrational patterns are usually different, leading to a poor mirror image. Even if a higher frequency absorption is excited, the fluorescence spectrum will only display one emission band, even though a molecule's electronic absorption spectrum may contain multiple absorption bands that correspond to multiple transitions from the ground state to the different excited singlet states.The reason for this is that the emission cannot compete with the radiation-less deactivation of the upper state due to the quick internal conversion and vibrational relaxation from the higher excited states. Only the lowest excited singlet state can produce fluorescence.

3.8. FLUORESCENCE

Fluorescence from multiple chemical species is almost usually indicated by the presence of multiple fluorescence bands in a fluorescence spectrum(Stich et al., 2010). Multiple fluorescence bands are frequently caused by contaminants. However, the many species that cause phantom fluorescences are occasionally produced by extremely quick physicochemical or chemical processes like isomerism and proton transfer. The effects on the position of the fluorescence band of substituents and the size of the aromatic system are qualitatively similar to those on the lowest frequency absorption band because fluorescence starts from the lowest excited singlet state and ends in the ground state.Therefore, when the size of the aromatic system increases, the energy gap (ground and excited states) diminishes in magnitude, which leads to a decrease in fluorescence frequencies. In their lowest excited singlet states, aromatic rings with single substituents such -OH, $-NH_2$, or -CHO are extensively conjugated than in their ground state. Regardless of whether the substituent is electron donating or electron receiving, the exocyclic conjugation stabilizes the excited state in comparison to the ground state, resulting in longer wavelengths of absorption and fluorescence than the unsubstituted aromatic compounds. Compared to strongly conjugating substituents, those that do not significantly conjugate with the aromatic ring have a far smaller effect on the excited state and, as a result, create smaller spectrum shifts. The shift of the fluorescence spectrum in relation to the parent hydrocarbon tends to approximate the vector sum of the shifting effects of each substituent when an aromatic ring contains multiple substituents. Molecular geometry, environmental factors like solvation or crystal habit, and electrical effects would need to be considered in a correct treatment of substituent effects. Not all potentially fluorescent molecules will fluoresce back to the ground state because fluorescence competes with internal conversion for the deactivation of the lowest excited singlet state. Due to the spin forbiddenness or low probability of fluorescence, the radiative transition from the lowest triplet state to the ground singlet state, known as phosphorescence, has a longer half-life.

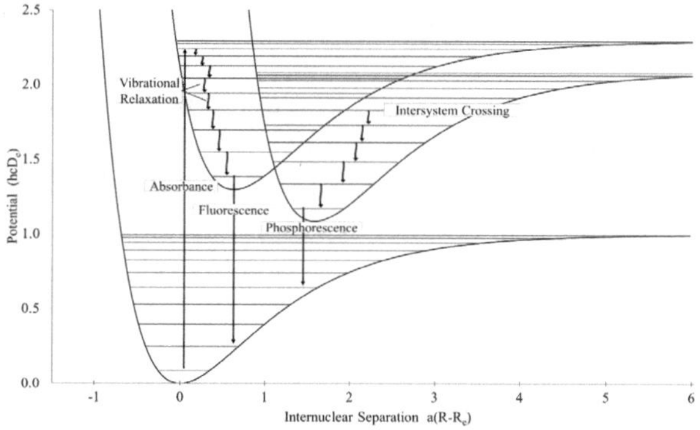

Figure 3.3. Phosphorescence.

Source: https://upload.wikimedia.org/wikipedia/commons/2/27/Electronic_Processes_Involving_Light.png, Creative Commons License.

3.9. PHOSPHORESCENCE

The most likely compounds to exhibit phosphorescence, (Figure 3.3)like fluorescence, are those with limited vibrational freedom, such as aromatic molecules and their derivatives. The majority of aromatic compounds lack phosphorescence. Triplet states are long-lived (10^{-5}s to several seconds) that phosphorescence and triplet-to-singlet intersystem crossing, as well as chemical and physical processes in solution, effectively compete for the deactivation of the lowest excited triplet state. The observation of phosphorescence in fluid media is always prevented by collisional deactivation by solvent molecules, quenching, photochemical reactions, and energy transfer mechanisms, with the exception of the short-lived phosphorescences. Instead, at liquid nitrogen temperature, phosphorescence is typically studied in the glassy state.When aromatic particles are adsorbed on filter paper, sodium acetate, and a few other substrates, as well as in ambient temperature solutions where the phosphorescing species is trapped in a micelle or a large enveloping molecule like a cyclodextrin, weak phosphorescence has occasionally been seen at room temperature. The lowest vibrational level of the lowest triplet state is where phosphorescence begins, and it can end in any of the ground state vibrational levels. The ground state vibrational structure is represented by the phosphorescence band structure. The 0–0 band of phosphorescence is often the lowest frequency vibrational characteristic of the phosphorescence spectrum(Itoh, 2012).The 0–0 band of phosphorescence always occurs at frequencies significantly lower than the 0–0 absorption band, in contrast to the 0–0 band of fluorescence, which may occur at the same frequency as the 0–0 band of the lowest frequency absorption maximum.This is a result of the lower repulsive energy in the triplet state, which causes the lowest triplet state to always lie below the lowest excited singlet state.

Similar to absorption and fluorescence, the energy at which an aromatic system phosphoresces decreases with its size. Phosphorescence can be distinguished from fluorescence by its longer decay time, its noncoincidence with the longest wavelength absorption band, even in low-temperature rigid matrices, and its nearly constant absorption of the phosphorescence emission at room temperature in fluid solution. If the molecule exhibits fluorescence at the temperature of measurement, the frequency of the extraneous band coincides with the fluorescence frequency, but its decay time is comparable to that of the phosphorescence. This is because the second band has a higher frequency than the phosphorescence. When a molecule in the lowest excited triplet state is thermally excited back to the lowest excited singlet state, fluorescence results. This phenomenon is called delayed fluorescence (Goltsev et al., 2009). Heating the sample or using statistical fluctuations in the triplet state thermal energy are two ways to achieve thermal excitation in solid media. When two triplet molecules collide in viscous fluids with considerable mobility, one is thermally excited to the lowest excited singlet while the other is converted to the ground state without radiation. Only specific molecules with a very small energy gap between the lowest excited singlet and triplet states exhibit the former process, which is known as E-type delayed fluorescence (eosin). This makes thermal repopulation of the lowest excited singlet state likely. Because it was initially seen in pyrene, the latter mechanism is known as P-type delayed fluorescence(Figure 3.4). It can occur even in molecules with significant energy differences between the lowest excited singlet and triplet states. The energy of the triplet state, which is typically more than sufficient to move a single triplet state molecule to the lowest excited singlet state, represents the total energy available.

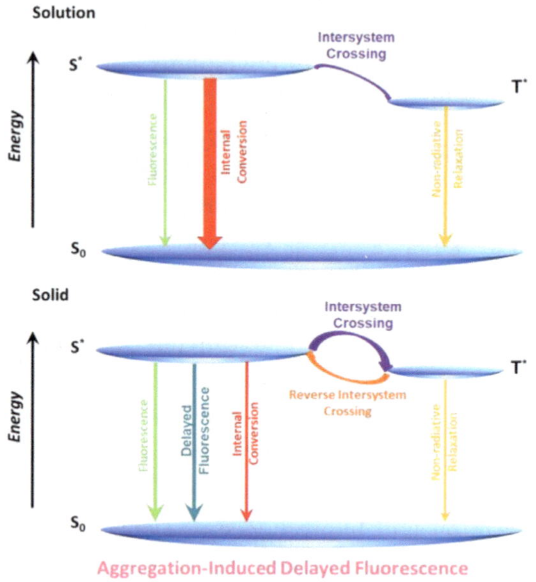

Figure 3.4. Delayed fluorescence.

Source: Yu, M., Huang, R., Guo, J., Zhao, Z., & Tang, B. Z. (2020). Promising applications of aggregation-induced emission luminogens in organic optoelectronic devices. *PhotoniX, 1*(1), 11. https://doi.org/10.1186/s43074-020-00012-y, Creative Commons License.

High concentrations of the absorbing particle are necessary because one P-type delayed fluorescence event requires two triplet molecules. Because two photons of exciting light must be absorbed in order to produce one event, P-type delayed fluorescence is a biphotonic process. As a result, the square of the exciting light intensity determines the intensity of P-type delayed fluorescence. E-type delayed fluorescence is monophotonic, and its intensity changes linearly with the first power of the exciting light intensity(Ziessel et al., 2013).As an analytical technique, phosphorescence spectroscopy has been developed, while delayed fluorescence spectroscopy has been developed easily. The main issues with their frequent use are related to reproducible sampling. Low emission yields from the analytes and high background luminescence from the substrates limit room temperature phosphorimetry, and the narrow bore tubes that must be used as sample containers for very low temperature spectral measurements are challenging to fill and reposition. The surroundings of potentially absorbing and fluorescing molecules may have a significant impact on the spectral position, intensity, and contours of molecular electronic spectra, even though these characteristics are primarily determined by molecular electronic structure and geometry. Solvation or packing into a crystal unit cell may change the orientation of substituent groups and, consequently, their involvement in an aromatic electronics structure. Spectra, particularly fluorescence spectra, may be substantially changed even in cases when the interactions with the environment are less obvious. The rates at which processes compete with light ions for the deactivation of the lowest excited singlet state and the fluorescence rate both affect the quantum yield of fluorescence. The rate of fluorescence may be significantly impacted by the solvent selection, but the rates of processes that compete with fluorescence are frequently more affected, and the choice may even determine whether fluorescence can be seen at all. Chemical processes that were kinetically or thermodynamically impossible in isolated molecules without solvent frequently take place in ground and electronically excited states in strongly interacting solvents. In order to explain spectrum behavior in condensed media, these interactions must be taken into consideration as they frequently have significant effects on the observable electronic spectra.

3.10. SOLVENTS

The main method that solvents affect electronic spectra is through their electrostatic characteristics. Depending on the solvent and solute molecular structures, they may interact with solute molecules via hydrogen bonds, dipole-dipole, dipole-induced dipole, or induced dipole-induced dipole. The dipole-dipole interaction frequently enhances hydrogen bonding when it does occur as a means of solvation. A hydrogen-bond donor solvent is one that contains hydrogen atoms bound to electronegative atoms that are capable of hydrogen bonding, such as oxygen, nitrogen, or Sulphur. A hydrogen-bond acceptor solvent is one that contains elements with unshared valence-shell electron pairs, like nitrogen and oxygen. A hydrogen-bond donor solvent acts as a very weak Bronsted acid by partially giving a proton to the solute molecule basic site or unshared valence electron pair. A hydrogen-bond acceptor solvent partially accepts a proton from the solute molecule by acting as a very weak Bronsted base. The sites of these kinds of spectra are greatly influenced by hydrogen-bonding solvents because unshared valence electron pairs are involved in giving the optical electron in $n \rightarrow \pi^*$ and intramolecular charge-transfer transitions. Solvent polarity has the greatest impact on $\pi \rightarrow \pi^*$ and intramolecular charge-transfer transitions of functionally substituted polar aromatic compounds due to the significant dipole moment shifts that accompany electronic reorganisation. Electronic spectra are frequently influenced by both dipolar and hydrogen bonding processes. Light is absorbed by a solute molecule with a permanent dipole moment that can form a hydrogen bond in a solvent with a high dielectric strength and both donor and acceptor hydrogen bonding qualities. The molecule will be solvated in the ground electronic state so that the solvent molecule positive poles point in the direction of the solute dipole negative ends and the solvent molecule negative poles point in the direction of the solute dipole positive ends. It is possible for the solvent molecule positively polarized hydrogen atoms to face unshared valence electron pairs on the solute, whereas the solute acidic hydrogen atoms may interact with unshared valence electron pairs on the solvent molecules.

The most stable solvent structure for the solute ground-state electronic distribution will be the solvent cage surrounding it(Leutwyler and Boesiger, 2002). The solute electronic distribution is changed by light absorption, resulting in a difference between the electronically excited molecule electronic dipole moment and that of the ground state molecule. The absorption process is quick that it ends with the electronically stimulated molecule still in the ground electronic state equilibrium geometry and encircled by the ground electronic state equilibrium solvent cage.There are stronger electrostatic forces if the solute molecule becomes more polar in the excited state. Contact with the polar solvent stabilizes the electronically excited state. Assuming that the shape of the molecule remains unchanged after excitation, a first approximation states that the energy of the electronically excited state will decrease with increasing solvent polarity.

This type of behavior, which tends to cause the absorption band to shift to longer wavelengths as solvent polarity increases, is seen in most $\pi \rightarrow \pi^*$ and intramolecular charge-transfer transitions.If the electronic dipole moment in the electronically excited state is smaller than in the ground state, the absorption spectra will tend to shift to shorter wavelengths as solvent polarity increases. The ground state will become more stable as solvent polarity increases, but the electronically stimulated state will become less stable. The solvent hydrogen-bond donor characteristics typically have a greater impact on absorptive transitions of the $\pi \rightarrow \pi^*$ type than does solvent polarity in general. When a solvent hydrogen atom coordinates a nonbonding pair on a solute molecule, the hydrogen-bonding interaction reduces the energy of both the ground state and the solute's n, π^* state. The ground state is more stabilized since it has two electrons in the nonbonding orbital, whereas the excited state only has one. As the solvent hydrogen-bond donor capacity increases, the energies of $n \rightarrow \pi^*$ absorptions rise. The positive end of the dipole of a polar solvent can also raise the energy of a $n \rightarrow \pi^*$ absorption band, but to a lower extent. Intramolecular charge-transfer absorption spectra are impacted by hydrogen-bonding solvents. By adding a partial positive charge to the functional group, hydrogen-bond donor solvents that interact with unshared valence electron pairs on functional groups that are charge-transfer acceptors in the excited state improve charge transfer.

As the solvent hydrogen-bond donor capacity increases, the absorption spectra tend to move to lower energies because of this interaction, which stabilizes the charge-transfer excited state in relation to the ground state. When a solvent interacts with unshared valence electron pairs on functional groups that are charge-transfer electron donors in the excited state, such as -OH or -NH_2, it typically results in shifts to higher energy(Mo et al., 2022).When hydrogen atoms are solvated on functional groups that are charge-transfer electron donors in the excited state, hydrogen-bond acceptor solvents typically result in shifts to longer wavelengths. This is because the positively charged proton partially withdraws from the functional group, allowing the electronic charge to move away from the functional group. By leaving a residual negative charge on the functional group, the solvation of hydrogen atoms on functional groups that are charge-transfer acceptors in the excited state prevents charge transfer. It is important to keep in mind that the Franck-Condon excited state may actually have more energy than the ground state if the geometrical configuration and solvent cage differ significantly from those in the equilibrium ground electronic state. This is because solvent polarity increases even when the absorbing molecule dipole moment increases upon excitation.

Therefore, the absorption spectrum moves to shorter wavelengths the more strongly the solvent interacts, irrespective of whether the absorbing molecule becomes more or less polar in the excited state. Following excitation, the Franck-Condon solvent cage and the molecular geometry in potentially luminous compounds accommodate the excited molecule altered electronic distribution(Feng et al., 2023). Reorienting the solute dipoles around new centers of positive (+) and negative (–) charge in the excited molecule is a phase of the relaxation process. It also likely involves the strengthening, weakening, breaking, and creating of hydrogen bonds, as well as the rotation of functional groups and substituents from aromatic rings into or out of coplanarity. In solvents with viscosities similar to water at room temperature, solvent relaxation occurs roughly concurrently with rotational and vibrational relaxation, taking around 10^{-13}–10^{-1}1 s, due to the involvement of nuclear movements or motions(Bryant, 1978). When compared to the deactivation of the lowest excited singlet state, solvent relaxation is often quick. The excited solute molecule in a thermally relaxed solvent cage with a lower energy than the Franck-Condon excited state is the source of fluorescence. The solute molecule ground electronic state is where the fluorescent transition ends, but due to the electronic transition speed, the molecule reaches the ground state while still having the equilibrium solvent cage and equilibrium geometry of the electronically excited state. The solute molecule eventually returns to the ground-state equilibrium solvent cage and geometry after rapid solvent and geometrical relaxation. Fluorescence frequently occurs at much longer wavelengths than would be expected based only on electrostatic stabilization or vibrational relaxation because the solvent-equilibrated excited state has a lower energy than the Franck-Condon excited state and the Franck-Condon ground state has a higher energy than the solvent-equilibrated ground state.

This explains why, particularly in solvents with strong interactions, the 0–0 bands of fluorescence and absorption frequently do not coincide. The electrostatic effects of hydrogen bonds and solvent polarity on fluorescence spectra are qualitatively comparable to those on absorption spectra. However, the relaxation processes that follow absorption and fluorescence always cause fluorescence to shift to longer wavelengths as solvent polarity and hydrogen-bonding capacity increase. The emission spectra are either pushed to longer wavelengths or unshifted by increasing the solvent proticity, while the absorption spectra are obviously shifted to shorter wavelengths by the same solvents, even in the rare instances of $\pi^* \rightarrow$ n fluorescence. By freezing the test solution, it is possible to illustrate the significance of solvent and geometric relaxation in determining spectrum changes. In order to prevent solvent cage and functional group relaxation after excitation, freezing locks the fluorescing molecule into the ground-state equilibrium solvent cage and typically the ground-state equilibrium as well.

The displacement of the fluorescence spectrum from the absorption spectrum in the rigid phase is smaller than in the fluid when comparing the fluorescence and absorption spectra obtained in fluid media and frozen solutions(Glasbeek and Zhang, 2004).These shifts to shorter wavelengths that are typically seen when small molecular fluorophores are attached to macromolecules may potentially be caused by this blocking of relaxational processes. The way the ground and Franck-Condon excited singlet states of the solute interact with the solvent determines the molar absorptivities of the different absorptive transitions, which in turn determines the intensities of absorption spectra in different solvents.In turn, molecule structure affects the molar absorptivities. The degree of electrostatic interaction between the functional groups and the aromatic ring are affected by the strong interaction of the functional groups on the aromatic rings with the solvent, particularly when charge transfer between the aromatic rings and functional groups takes place in the excited state.As in the case of arylamines in strongly protic solvents or carboxylic acids in hydrogen-bond acceptor solvents, solvation that disrupts the electronic contact between the ring and functional group tends to reduce the molar absorptivity. The solvent has minimal effect on the molar absorptivity; unless the solvent influence is so strong as to totally prevent intramolecular charge transfer in the absorptive process. The polarity of the solvent and hydrogen-bonding characteristics have a significant impact on the fluorescence spectrum intensities. The effects of solvent perturbations on the electronic absorption spectra are analogous to these solvent-induced sensitivities, which are partially caused by changes in the radiative transition probability (fluorescence) caused by solvent-induced enhancement or inhibition of intramolecular charge transfer in the electronic transitions of luminescing molecules.

3.11. SOLVENT MEDIATED MODIFICATIONS

Solvent-induced modifications to the rates and mechanisms of deactivation of the electronically excited singlet states that give rise to fluorescence, however, have the major effects on the intensities of the luminescence transitions. Increases in solvent hydrogen-bonding capacity often result in a fall in fluorescence quantum yields in many molecules whose lowest excited singlet states are of the π, π *, or intramolecular charge-transfer types. Fluorescence may be seen when the solution freezes in some solvents when it does not occur in fluid circumstances. These findings imply that the low fluorescence yields in activating solvents are caused by hydrogen-bonding interactions in the equilibrium excited-state solvent cage. Reduced solvent activating strength leads to weaker contacts in both the ground and excited states, whereas freezing the solvent stops solvent relaxing in the excited state and restricts the solute to the ground-state solvent cage. It is possible that the solvent-solvent association found in solvents with strong hydrogen bonds offers a mechanism to deactivate the fluorophore attached to the solvent without using radiation by using the solvent quasi-lattice vibrations. Internal conversion to the solute-solvent supramolecule would be facilitated by the solvent vibrations. In certain instances, the solvent hydrogen-bonding characteristics are the only factors influencing fluorescence efficiency. Dependency on both solvent qualities is seen in some situations, whereas in others, it varies with solvent polarity. Sometimes the effect of hydrogen-bonding strength or solvent polarity on fluorescence efficiency is not constant, and changes in one of these solvent characteristics can cause fluorescence to rise and fall. The external heavy-atom effect is one kind of solvent effect that has a significant impact on fluorescence intensity but has minimal effect on transition frequency. The high nuclear charge of the heavy atoms in solvent molecules of the solute molecule primary solvent cage results in a strong interaction between the solute spin and orbital electronic angular momenta in solvents whose molecules contain atoms of high atomic number, such as iodine. The loss of the total molecule spins angular momentum, a molecular feature (quantum number), is caused by the interaction of the spin and orbital electron movements, which is very weak in the absence of heavy atoms. Phosphorescence, singlet-triplet intersystem crossing from the lowest excited singlet state, triplet-singlet intersystem crossing from the lowest triplet to the ground state, and singlet ➔ triplet absorptive transitions (Figure 3.5)are more likely as a result of the partial removal of the selection rule that prohibits changes of spin angular momentum in electronic transitions.

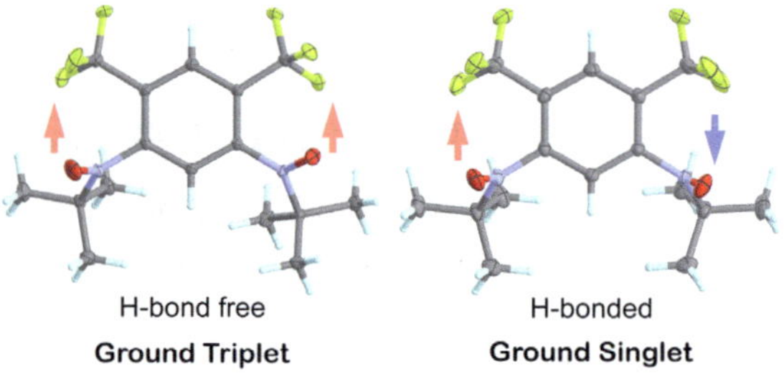

Figure 3.5. Singlet triplet absorptive transitions.

Source: Haga, N., & Ishida, T. (2024). A triplet/singlet ground-state switch via the steric inhibition of conjugation in 4,6-bis(trifluoromethyl)-1,3-phenylene bisnitroxide. *Molecules, 29*(1), Article 1. https://doi.org/10.3390/molecules29010070, Creative Commons License.

As a result, weak absorption bands at frequencies lower than the lowest frequency singlet-singlet absorption bands appear in certain aromatic compounds immersed in heavy-atom liquids.Since the lowest triplet state has less energy than the singlet state with the same electronic orbital structure, the lowest singlet ➔ triplet absorption bands occur at lower frequencies than the lowest singlet ➔ singlet absorption bands. In heavy-atom solvents, molecules that fluoresce in common solvents such as water, alcohols, and hydrocarbons typically glow less intensely or may even have their fluorescences totally extinguished (Wehry, 1990).This is due to the heavy-atom solvent higher singlet-triplet intersystem crossing probability (rate), which causes the population of the lowest triplet state to grow at a faster pace than that of the lowest excited singlet state. Fewer molecules can fluoresce at any given moment. Phosphorescence is favored by the large population of the triplet state created by the external heavy-atom effect, and phosphorescence yields in rigid matrices are frequently increased in heavy-atom solvents. In the least excited singlet state, electron-donating groups have a tendency to become more acidic or less basic, and increases in acidity upon excitation are linked to a spectral shift to a longer wavelength upon dissociation or a shorter wavelength upon protonation.

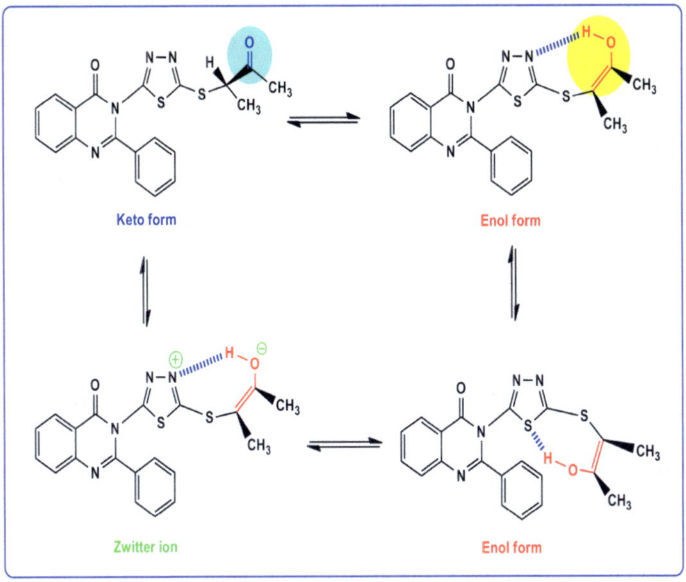

Figure 3.6. Tautomerization.

Source: Mohammed, S. J., Salih, A. K., Rashid, M. A. M., Omer, K. M., &Abdalkarim, K. A. (2020). Synthesis, spectroscopic studies and keto-enol tautomerism of novel 1,3,4-thiadiazole derivative containing 3-mercaptobutan-2-one and quinazolin-4-one moieties. *Molecules, 25*(22), Article 22. https://doi.org/10.3390/molecules25225441, Creative Commons License.

3.12. TAUTOMERIZATION

When a polyfunctional molecule has at least one acidic electron-donating group and one basic electron-accepting group, the donor group may become very acidic. The donor group may simultaneously lose a proton and the acceptor group may receive one. In the excited state, this is a tautomerization or isomerization (phototautomerism)(Waluk, 2006)(Figure 3.6). The molecule that underwent phototautomerization does not have a net gain or loss of proton. Since tautomerization occurs after the absorption process, phototautomerization is not shown in the electronic absorption spectrum. However, the anomalously long wavelength fluorescence spectrum will clearly show phototautomerization if it takes place during the span of the lowest excited singlet state. In comparison to the fluorescence of the unphototautomerized molecule, the shift to long wavelength of the phototautomer fluorescence will roughly equal the sum of the spectral shifts caused by protonation and dissociation separately. However, a wide frequency gap between the fluorescence maximum and the long wavelength absorption maximum is frequently a sign that phototautomerization has occurred in the excited singlet state. The opposite sequence of spectral shift of absorption and fluorescence shift with changing pH are frequently observed due to the frequent inversion of the order of protonation of functional groups in molecules that undergo phototautomerization.

The phototautomerization in the lowest excited singlet state is frequently indicated by a shift to a longer wavelength followed by a shift to a shorter wavelength of the absorption spectrum, as well as a shift to a shorter wavelength followed by a shift to a longer wavelength of the fluorescence spectrum with decreasing pH. The result of intramolecular or intermolecular proton transfer can be phototautomerism. The direct transfer of a proton from the electron donor group to the electron acceptor group is known as intramolecular phototautomerism. This process is frequently seen in polyfunctional molecules where the two groups are orientated orthogonally to each other. In the ground state, a proton of the donor group and an acceptor group often form a rather strong intramolecular hydrogen bond. It was found that salicylic acid fluoresced at very low frequencies in polar liquids. Salicylic acid showed two fluorescence bands in hydrocarbon solvents. One resembled the emission seen in ethanol and was at a long wavelength (440 run). The other one was at a shorter wavelength (340 nm), looked like the fluorescence of *o*-methoxybenzoic acid, a molecule with an electronic structure similar to salicylic acid but without a phenolic proton that could be transferred to the carboxyl group, and formed a mirror image with the long wavelength absorption band of salicylic acid.It was determined that the neutral molecule (N) was responsible for the absorption and short wavelength fluorescence of salicylic acid in hydrocarbon solvents, while the zwitterion (Z), which is formed as a result of proton transfer in the lowest excited singlet state, was responsible for the long wavelength fluorescence.

The process had to be intramolecular since salicylic acid phototautomerization took place in hydrocarbon solvents, which were unable to facilitate proton transfer. Investigations into the pH dependency of salicylic acid fluorescence revealed that phototautomerism was pH independent, suggesting that the activity was intramolecular even in water. It is known that compounds with widely separated donor and acceptor groups exhibit phototautomerism. In these situations, proton transfer needs to be multimolecular, involving solvent support for the acceptor group protonation and the donor group dissociation. Non-transition-metal ions, such as Zn(II), Al(III), and Ca(III), can coordinate aromatic ligands by causing positive polarization at the ligand coordination site(Aly et al., 2023). This results in spectral shifts that are comparable to those caused by protonating the ligand at each of the ligand coordination sites. Therefore, in ethanolic solutions, 8-hydroxyquinoline, which is colorless and fluoresces blue, turns yellow and fluoresces green when it reacts with a hydrogen ion or a non-transition metal ion (Alday et al., 2020). The 8-quinolinolate non-transition-metal complex and the 8-quinolinolium cation have slightly different low-frequency absorption maxima and fluorescence maxima. The absorption maxima is around 370 nm, whereas the fluorescence maxima is around 500 nm. This is due to the different polarizing capacities of the proton and the various non-transition-metal ions. Both the protonated ligand and the ligand complexed with the different non-transition-metal ions tend to have significantly different quantum yields of fluorescence. Due to spin-orbital coupling, which populates the lowest triplet state at the expense of the lowest excited singlet state, the quantum yield of fluorescence decreases as the metal ion atomic number increases. Electronic spectrum shifts resulting from complexing an aromatic ligand with transition-metal ions are often significantly larger than those resulting from complexing the identical ligand with non-transition-metal ions.

CHAPTER 4

LASER SPECTROSCOPY

4.1. INTRODUCTION

The unique characteristics of laser light sources are what set them apart from other types of light sources. These characteristics specifically include monochromaticity, spectral intensity, wavelength tunability (frequently over intervals), temporal and spatial coherence, and temporal fluctuation from continuous operation to ultrashort pulses. The laser differs from traditional or conventional light sources primarily due to its intensity within narrow spectral intervals. This is due to the relation $I_{signal} \propto N \ldots S_{\lambda} \ldots . I_{\lambda}$ can be used to describe how a system responds to a light stimulus. N represents the number of atoms or molecules involved in the process, S_{λ} includes the transition probability for the photon interaction, and I_{λ} is the light source spectrum intensity, matched to the transition bandwidth. Compared to regular light sources, the spectral intensity can be higher. This indicates that a higher signal-to-noise ratio is possible. It is simpler to investigate processes in which very few particles are involved. Photodissociation fragments, transient species, reaction intermediates, and others can be traced in chemistry, and signals at low particle densities can be recorded in analytical measurements, allowing us to examine individual particles of a species. High monochromaticity and laser tunability make it possible to do high-resolution spectrum observations. This means that individual quantum states can be chosen, even resolving quantum properties like hyperfine structure or targeting an isotope of a species. It also means that laser light excitation can be molecule-specific, which is very advantageous in chemical reaction kinetics or compositional analysis. Many applications for laser light are made possible by its excellent coherence qualities, such as time-delayed overlap between two light beams or focusing. High temporal coherence will enable the customization of short laser pulses, which can be used in regulated pump-probe experiments to investigate ultrafast processes in real time. Spatial coherence is a feature that can be used to separate individual atoms or molecules or to focus a laser beam to achieve high power densities (Cao et al., 2019).

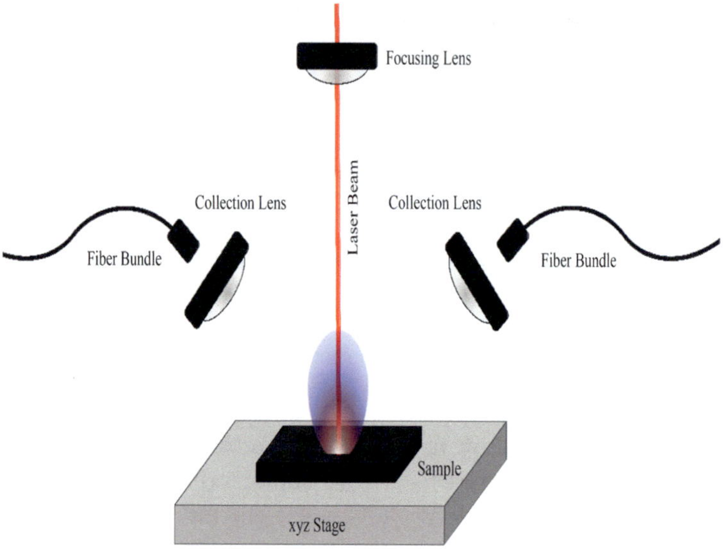

Figure 4.1. Laser spectroscopy.

Source: Stefas, D., Gyftokostas, N., Bellou, E., &Couris, S. (2019). Laser-induced breakdown spectroscopy assisted by machine learning for plastics/polymers identification. *Atoms, 7*(3), Article 3. https://doi.org/10.3390/atoms7030079, Creative Commons License.

Although light pulses as short as 1 μs are possible, lasers are generally used for any shorter pulse duration, including nano-, pico-, femto-, and attosecond. The field of ultrashort pulses with tp ~ 10^{-15}–10^{-18} s has allowed scientists to access chemical and physical processes that were previously only theoretically possible but lacked experimental validation. To fully comprehend chemical reactivity, one must fully understand an elementary chemical reaction that takes place in a single, short time event. Femtochemistry is created by the ability to study the intermediate steps in such a reaction both dynamically (time) and spectroscopically (energy). The imaging of electron quantum motion in both real time and space is an example of the utility of ultrashort attosecond pulses that goes beyond the basic chemistry. The state of matter and the degree of aggregation have a substantial impact on the spectroscopy and imaging of atoms, molecules, and chemical reactions of samples.

4.2. LASER SPECTROSCOPY

Laser spectroscopy (Figure 4.1) techniques have been developed to analyze processes in the gas phase, clusters, solutions, and surfaces (Demtröder, 2015). These techniques are tailored for specific scenarios. Over the past fifty years, many laser spectroscopic techniques have been created; yet, the fundamental principles of spectroscopy have remained largely same, with the main difference being the light source. Therefore, the core of spectroscopic techniques is photon absorption, emission, and scattering processes and their implications for the examined material.

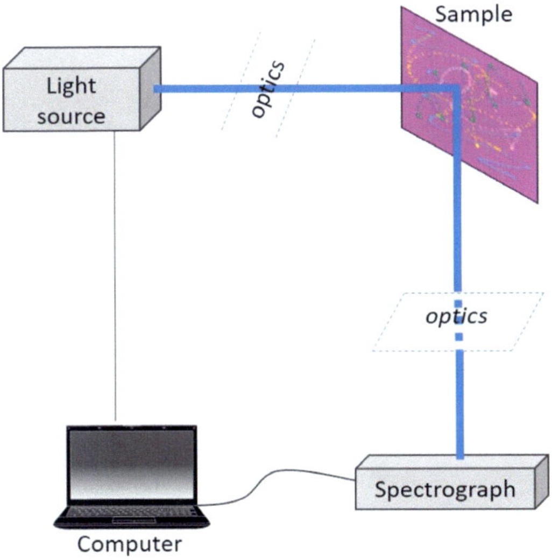

Figure 4.2. Laser-induced fluorescence spectroscopy (LIFS).

Source: Ghervase, L., &Cortea, I. M. (2023). Lighting up the heritage sciences: the past and future of laser-induced fluorescence spectroscopy in the field of cultural goods. *Chemosensors, 11*(2), Article 2. https://doi.org/10.3390/chemosensors11020100, Creative Commons License.

However, the laser made it feasible for researchers to take advantage of processes that are hard or impossible to achieve with traditional light sources. Femtosecond spectroscopy and nonlinear spectroscopic methods are important examples. In case of absorption spectroscopy, many methods have been transformed by the laser. When using tunable laser sources, the spectrum selectivity is typically provided by the laser itself, avoiding the need for a spectrometer. Because commercially available laser sources with extremely narrow bandwidth can resolve hyperfine and isotopic details that were previously impossible to resolve with standard spectrometers, extremely high resolution can be achieved. Compared to absorption spectroscopy, laser-induced fluorescence spectroscopy (LIFS)(Figure 4.2)is less widely used in commercial applications.

This is primarily because it is very challenging to quantify the signals that are observed, which limits its application in quantitative analysis, which is what commercial users aim for. However, LIFS is still frequently the preferred approach for a variety of scientific problems because access to certain quantum states is made possible by the laser narrow bandwidth. It enables extremely high sensitivities by fusing the selectivity of the absorption transition with the background-free observation of fluorescence photons, eventually engaging just one atom or molecule.

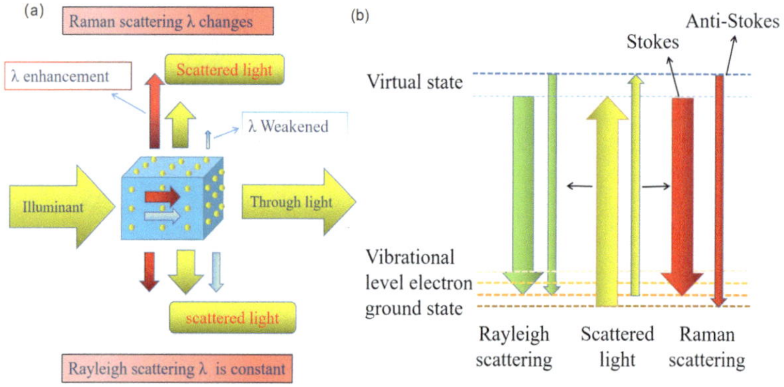

Figure 4.3. Raman spectroscopy(a and b)

Source: Li, L., Cao, X., Zhang, T., Wu, Q., Xiang, P., Shen, C., Zou, L., & Li, Q. (2022). Recent developments in surface-enhanced raman spectroscopy and its application in food analysis: alcoholic beverages as an example. *Foods, 11*(14), Article 14. https://doi.org/10.3390/foods11142165, Creative Commons License.

Raman spectroscopy (Figure 4.3)has largely advanced from the development of the laser, despite being restricted for a long period to use in research laboratories. Its universality, such as gaseous, liquid, and solid samples, can be probed, and its ability to support both qualitative and quantitative analysis of multispecies samples have made it one of the backbones of commercial analytical spectroscopy today. This is especially true for biological and pharmaceutical applications, where it has proven to be invaluable. Initially, the stimulated atoms may experience spontaneous emission; when connected to the original, stimulating (absorbed) photon, the light phase is randomly dispersed. Second, the transition is started by an incident photon, which causes a photon to be emitted as the atom changes from the excited to the ground state. This process is known as stimulated emission, and it was first suggested and accurately explained by Einstein. The stimulated emitted light has the same frequency and phase as the incident light, and that this process is the behind the laser emission. It should be noted that the two emission processes are fundamentally distinct from one another due to their random/equal phase conditions: spontaneous emission is not a coherent event, but the stimulated process is. Water may absorb in other parts of the spectrum even though it is transparent in the visible range. Actually, the way that water absorbs in the UV and IR is widely understood. The greenish-blue color of seawater is caused by an absorption in the red near 750 nm by water molecules; this should not be mistaken with the blue that is frequently seen in connection with the reflection of the blue sky. Various relaxation processes may occur when a photon is absorbed and a (organic) molecule is excited.

4.3. TRANSITIONS

Radiative and nonradiative transitions are the two categories into which these excited-state photophysical processes fall. The two radiative processes, phosphorescence and fluorescence, have the following primary characteristics. Usually starting at the lowest vibrational level of the lowest excited singlet state, fluorescence is a radiative transition between states of the same multiplicity. Phosphorescence is a spin-forbidden radiative transition between states with varying multiplicities, typically from T_1, the lowest triplet state, back into S_0 at its lowest vibrational level ($v = 0$)(Baryshnikov et al., 2017). Fluorescence is a faster process than phosphorescence; phosphorescence emission often occurs on timescales of 10^{-3}–10^{2} s, whereas fluorescence emission typically occurs in the picosecond to microsecond region, or 10^{-12}–10^{-6}s. This indicates that as soon as the stimulating radiation is turned off, spontaneous fluorescence emission stops. Phosphorescence, on the other hand, can exhibit emission for several seconds following the cessation of excitation. The existence of a distinct number of photophysical stages leading to both radiative processes is suggested by these varied timings. The initial photon absorption from the ground to the excited electronic state is shared by both processes. Since its creation in 1960, the laser has established itself as a fundamental component of imaging and spectroscopy, mostly because of its extraordinary technological advancements as a dependable and adaptable light source for a wide range of applications. The quantum interactions and processes that cause the population to invert between the upper and lower laser levels are crucial; these include the pumping of the laser-active state and the internal transitions in the active medium, some of which are undesired and interfere with the crucial laser transition. The actual laser cavity that, upon repeated passage through the laser-active medium, transforms the population inversion into useful laser radiation via stimulated emission; additional optical elements within the cavity, in addition to the actual mirror selection and cavity configuration, can be used to customize the laser output.

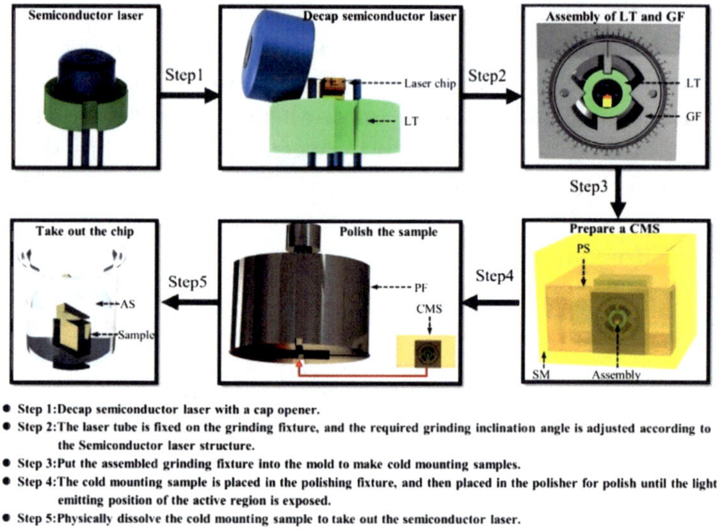

Figure 4.4. Semiconductor diode lasers.

Source: Sun, T., Qiao, L., & Xia, M. (2021). Effective failure analysis for packaged semiconductor lasers with a simple sample preparation and home-made PEM system. *Photonics, 8*(6), Article 6. https://doi.org/10.3390/photonics8060184, Creative Commons License.

The characteristics of the laser radiation, such as its polarization, pulse duration, beam profile, wavelength, line width, and light power/energy included in the beam, that are sent from the laser to the experiment site. Only the laser-active quantum characteristics and photon transition rates were considered in the conceptual approach of optical pumping and level populations. Feedback must be used to enhance some of the original spontaneous photons and employ the stimulated emission process to turn them into the practically useable laser light wave in order to take advantage of any population inversion between the potential laser levels. This requires setting up an optical cavity arrangement that will resonate and magnify light waves inside the cavity. A standing wave pattern is created that must satisfy a number of boundary criteria or conditions in both space and frequency. Two mirrors facing one another make up an optical cavity. Two plane mirrors create a Fabry-Perot cavity, which is the basic and often used arrangement (Bitarafan and DeCorby, 2017). Despite being essentially simple, flat mirrors are rarely utilized because, unless the cavity length is extremely small, they are challenging to align precisely enough to allow a standing wave pattern.

Semiconductor diode lasers (Figure 4.4)are an exception to this rule, as they have an inherent, precisely aligned plane-parallel cavity. Different resonator types result from the inclusion of curved mirrors in the configuration, both concave and convex. These are differentiated by the cavity length L, which is the distance between the focal lengths of the mirror(s), which have radii of curvature R_1 and R_2. Regardless of the cavity geometry, it is crucial to select one that keeps the oscillating wave stable, meaning it can sustain enough oscillations to boost the light wave to useful power levels. Only specific ranges of values for R_1, R_2, and L produce stable resonators. Any configuration is unstable if the stability parameter is less than zero or more than one. The cavities are partially stable if their parameter values match any of the stability region boundary values, such as the axes or the hyperbola lines. The resonator could become unstable with even a small change in cavity length or misalignment. Since the formalism was created assuming that the cavity is empty, the treatment of laser cavities is simple. This is not correct for a laser cavity since the cavity contains the gain medium as a minimal entity.

4.4. MIRRORS

The optical path length between the mirrors, or the effective lengths determined by the refractive indices, must therefore be considered rather than the physical distance between the mirrors, which is represented by the value of L. Many contemporary commercial laser systems have more mirrors than the two-mirror cavity; a variety of three- and four-mirror configurations are frequently seen, such as Z-shaped cavities and V-type and ring cavities. With the laser-active medium at the site, a confocal section is typically formed by a pair of curved mirrors; the rest of the cavity is made to be quasi-collimated using plane mirrors. Frequently, other components are positioned inside the cavity in addition to the gain medium. These could be spatial filters for transverse mode control or modulators for Q-switching and mode-locking in confocal sections of the cavity; other parts, such as filters, prisms, or diffraction gratings, generally need the beam going through them to be quasi-collimated. The level populations and cavity behavior were established on the idea that the steady state had been attained, meaning that no longer was there any temporal change in any of the features. The progression toward the steady state is time-dependent, even while steady-state conditions are certainly desirable in general, one desires consistent, non-varying laser output. Specifically, while turning on the laser or with pulsed lasers, temporal variations are anticipated. It is assumed that the entire system operates in the small-signal gain domain. This indicates that there is never a large population in the pumping level 4 and the lower laser level 2, i.e., $N4 \approx N2 \approx 0$, and the population in the ground state is not exhausted and is therefore roughly constant.

4.5. DOPPLER EFFECT

The spectral transition lines show broadening linked to the Doppler effect when the emitting particles (atoms or molecules) are free to travel through space, such as in a gas. The Doppler shifts of the individual emitters produce a statistical broadening profile due to the cumulative effect of the emitting particles moving at varying velocities. An observer (z) sees radiation with frequency $v = v_0\left(1 \pm v_z / c\right)$ when a particle emitting with frequency v_0 moves with velocity v (Misra and Brodin, 2022). The $+$ and $-$ signs correspond to approaching and retreating particles, respectively. The light intensity distribution follows the Gaussian distribution function, assuming that the particle ensemble obeys the thermal Maxwell–Boltzmann distribution. The envelope of emission processes from individual (spontaneous) emitters with Lorentz line profiles, dispersed based on their thermal statistics for individual frequency components, is first represented by the observed Gaussian function. The Doppler width δv_D is proportional to the transition frequency v_0. This indicates that Doppler broadening is especially significant for short wavelengths in the visible and ultraviolet regions of the spectrum; for instance, at room temperature ($T \cong 300K$), $\delta v_D \leq 1GHz$ is usually found, however light atoms and molecules may experience it up to around 10–30 GHz(Demtröder, 2010). Doppler broadening is significantly less for IR spectral lines, and the Doppler width may be equal or even less than the emitter native linewidth. Voigt profile is commonly seen in laser spectroscopy, is produced by the convolution of the Gaussian and Lorentzian functions, which have similar widths. Only gases are affected by the Doppler effect. Doppler broadening is not present in atoms or molecules implanted in crystalline or amorphous solids because they are locked into place. In a laser, the extrinsic characteristics of the cavity must be considered, even though the Lorentzian and Gaussian line profiles previously mentioned may be considered intrinsic to the quantum structure of the emitting or absorbing particle. It is obvious that this laser light will have a finite line width even if, the laser is only oscillating in single frequency mode. Phase noise or quantum noise related to spontaneous emission in any free-running laser oscillator is the cause of this. It can be demonstrated that the instantaneous frequency noise is known as white noise since the spectrum itself has a Lorentzian structure and its power spectral density remains constant.

4.6. QUANTUM

Schawlow and Townes in 1958 determined the fundamental (quantum) limit for the width of such a laser mode even before the laser was experimentally demonstrated; this limit is commonly referred to as the Schawlow-Townes equation. The Schawlow-Townes limit is rarely attained in practical laser systems due to the presence of several noise sources, such as temperature fluctuations and mechanical vibrations, which are challenging to suppress, in addition to quantum noise(Pollnau and Eichhorn, 2020). While creating a laser with an ideal thin line width, some general assumptions must be made. Although a long laser resonator may theoretically indicate a narrow Schawlow-Townes line width, mechanical stability is typically challenging to accomplish, and consistently stable single-frequency operation without mode hops is rarely possible. Free-running single-frequency semiconductor laser diodes often have line widths of a few megahertz. External-cavity diode laser setups can significantly reduce this width. Some researchers measured laser line widths around <100 kHz with a relatively simple but well-designed setup, which was still a factor 2 >their calculated Schawlow-Townes limit, indicating that technical noise still played a significant role in the linewidth. When stabilizing the laser system to a reference cavity, diode laser line widths in the subhertz zone have been demonstrated with instrumental efforts. It is evident that the laser will only oscillate on a single frequency in very few instances when comparing the width of the several broadening mechanisms with the (longitudinal) frequency mode spacing. Argon ion (Ar^+) lasers have a Doppler-broadened line width of $\Delta \nu G \cong$ 3–5 GHz due to the high discharge temperature. The cavity lengths of typical Ar^+ lasers range from 15 to 250 cm for low-power and high-power systems, respectively. The longitudinal frequency mode spacing value for the short cavity is determined to be δv_q = c/2L = 1 GHz(Song and Lin, 2024). This indicates that a wide range of modes may oscillate and contribute to the laser output by fitting under the (homogeneous) laser line profile. The laser gain bandwidth is sometimes defined as the frequency range over which laser oscillation is feasible above threshold. The laser output spectrum is influenced by the modes that oscillate above threshold while staying inside the parameters of the gain profile. Because gain can be defined in a number of ways, the precise meaning of the term gain bandwidth may not always be immediately apparent. The laser medium gain bandwidth is a crucial factor to take into account in a variety of scenarios, such as the following. For single-frequency operation, a narrow gain bandwidth is ideal. To achieve high gain efficiency, a limited gain bandwidth is required. A wide gain bandwidth is essential for a mode-locked laser to generate ultrashort pulses. Inadequate gain bandwidth may restrict the range for wavelength tuning. Depending on the laser medium gain bandwidth, a laser may oscillate on multiple frequency modes. The occurrence of multimode emission is actually more common. However, it is preferable for the laser to only function in one mode for a large number of spectroscopy and imaging applications. Pumping only the volume covered by the axial (00) modes or adding an appropriate, restricting aperture into the cavity are the usual ways to prevent the stimulation of higher-order transverse modes (TEMnm). When the gain bandwidth exceeds the longitudinal mode spacing, several longitudinal (axial) modes TEM_{00q} are still activated or excited. There are other techniques to decrease the number of oscillation modes, however the following approaches are the most commonly used. By increasing the axial mode spacing, i.e., by using very short laser cavities, which is impractical for many lasers or by introducing an appropriate intracavity filter, such as an etalon, which forms a coupled-cavity configuration; or by decreasing the gain bandwidth above threshold, which is feasible, for example, by carefully reducing the pumping but which consequently limits the available output power. Note that external-cavity feedback is utilized to execute oscillation on a specific, single frequency in place of intracavity filtering. It should be noted that inhomogeneous media can make single-frequency operation challenging due to gain saturation brought on by spatial hole burning in a standing–wave cavity. For this reason, unidirectional ring lasers are frequently used to achieve dependable and clean single-frequency operation. It should be evident from the rate equations alone that the level population / time evolution and the laser activity are always present. Relaxation oscillations rather than a smooth transition to laser activity occur once the pumping process is turned on. The system settles into an equilibrium state of level populations and laser output when the proper conditions are met. Lasers are categorized into two classes: continuous wave (CW) and pulsed lasers(Michalik et al., 2021). The latter is further separated into distinct groups based on the actual pulse duration, which is typically linked to various technical processes in the laser cavity.

4.7. PULSES

The classification of slow pulses, which are produced by modulating a CW laser output, short pulses, which are typically in the nanosecond regime and achieved by Q-switching, and ultrashort pulses, which are in the picosecond, femtosecond, or even sub-femtosecond regimes and are primarily produced by mode-locking techniques, is frequently used. When a laser operates in CW mode, it is continually pumped with enough force to maintain population inversion and continuously extract laser light from the cavity. There may be one or more resonator modes where this emission takes place. Even though there are many CW lasers on the market today, it should be noted that CW operation is not required to realize that CW operation is difficult to achieve in lasers with low-gain media transitions, where either a high pump power is needed or the achievable laser output power is low; additionally, self-terminating laser transitions are not suitable for CW operation at all. Although CW laser radiation is good, amplitude-modulated light is preferable for many spectroscopic and imaging applications.

Figure 4.5. Electro-optic modulator (EOM).

Source: https://upload.wikimedia.org/wikipedia/commons/f/f1/Electro-optic-modulator-linos.jpg, Creative Commons License.

To minimize undesired contributions from light, the detection techniques of lock-in and boxcar integration are commonly employed; these require appropriately modulated laser light, or stroboscopic pulsing is essential in certain laser imaging applications. Analog and digital modulation are the two primary ways that light is applied. The light intensity is switched between "on" and "off" or between preset intensity levels in the latter case; in the former, the light amplitude is modulated in accordance with a specific waveform, such as a smooth, sine function, though even arbitrary waveforms may have advantages. External to the laser cavity, a mechanical (spinning) light beam chopper is likely the application of light amplitude modulation. Generations of laser spectroscopists have employed this method and continue to do so. Despite its simple methodology, the technology has certain drawbacks. For the short-duration end of the spectrum, the rise and fall times of the modulation are constrained by the speed at which the chopper blade moves over the spatially expanded laser beam. Repetition rates and pulse durations are often restricted to the 10^{-3}–10^{-5} s range. The refractive modulators, which are based on electro-optic (EO) or acousto-optic (AO) effects, can take the place of mechanical choppers when higher modulation frequencies, into the hundreds of kilohertz or megahertz regime, are required.

4.8. ELECTRO-OPTIC MODULATOR(EOM) AND AOM

A device called an EOM(Figure 4.5)uses a nonlinear crystal whose refractive index is altered by an applied electric field; this phenomenon is sometimes referred to as the Pockels effect(Kawanishi, 2022).

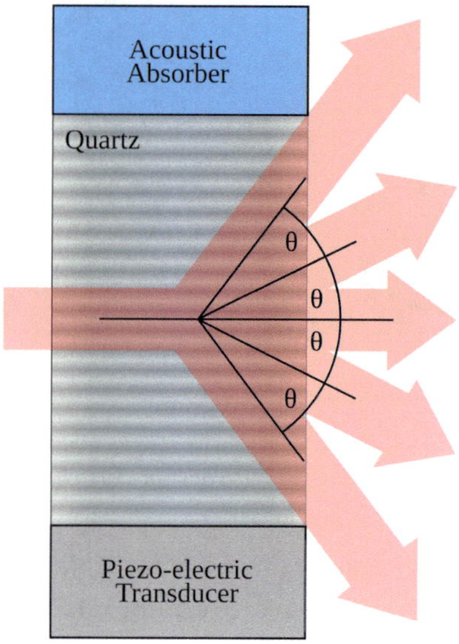

Figure 4.6. Acousto-optic modulator (AOM).

Source:https://en.wikipedia.org/wiki/Acousto-optic_modulator#/media/File:Acousto-optic_Modulator-en.svg, Creative Commons License.

EOMs can be used to regulate a laser beam polarization. One small disadvantage is that EOMs, also known as Pockels cells, typically need hundreds or even thousands of volts, necessitating the use of a high-voltage amplifier, which is not necessarily simple to design for high modulation frequencies. Pockels cells can function as light amplitude modulators when paired with polarizers. The polarization state is altered by the Pockels cell, and the polarizer then transforms this into a change in the transmitted laser-light amplitude and power.

An AOM(Figure 4.6)is a device that uses a sound wave oscillating mechanical pressure to alter the refractive index(Tran, 2012). With an electrical drive signal, EOMs can be utilized to control a laser beam power, frequency, or spatial direction. An AOM active component is a transparent crystal that has a piezoelectric transducer attached to it. The transducer causes a sound wave to be excited inside the crystal. AOMs are also referred to as Bragg cells because the sound wave creates a traveling periodic refractive index grating at which light undergoes Bragg diffraction(Figure 4.7).

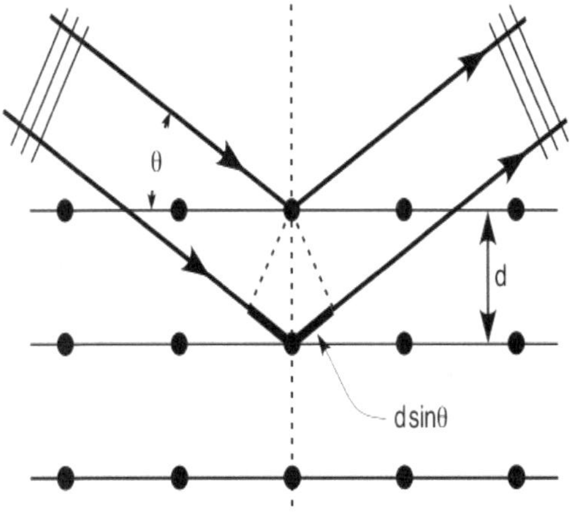

Figure 4.7. Bragg diffraction.

Source: https://upload.wikimedia.org/wikipedia/commons/0/0a/Bragg_diffraction.png, Creative Commons License.

Over 50% of the optical power can be diffracted at high acoustic powers. Modulation frequencies up to around 100 MHz can be provided by devices based on any many approaches. EOMs must be employed if extremely high contrast ratios between the minimum and highest transmitted light amplitudes are needed. These modulators can also be included into the laser cavity to modulate the laser action itself, as opposed to being used externally, that is, to modulate the laser beam. Unless there are limited options, cavity-internal EOM/AOM modulation devices are mostly employed for short-pulse and ultrashort-pulse generation rather for modulating a CW laser. Modulation is accomplished by changing the pump rate of the laser medium, driving the laser to higher or lower efficiency, or above or below the laser threshold, in order to modify the laser activity itself, whether by analog or digital wave forms. However, due to the laser dynamics, careful design considerations must be made while attempting to optimize the modulation period in terms of duty cycle, overshoot and ringing, extinction ratio, and rise and fall duration. In semiconductor diode and quantum cascade lasers, which are directly pumped by the injection current and hence well suited for electrical signal wave modulation, or in lasers that are pumped by diode lasers, laser modulation is utilized. Any system that does not function in the CW mode can be referred to as a pulsed laser since its output pulses have a specific repetition rate and a certain duration. One common confusion in this regard is that modulated and pulsed lasers can be treated somewhat interchangeably. A modulated laser changes the laser output between the "low" and "high" levels, independent of the modulation speed, whereas a pulsed laser yields a varied burst of light energy based on the actual pulsing conditions. Peak pulse power levels that significantly exceed the (CW) steady-state power level can be produced for a short period of time by appropriately adjusting the pulse behavior.

4.9. PULSE WIDTH AND Q SWITCHING

The pulse width or duration is one of the crucial characteristics in the description of the laser pulses that are generated. The FWHM parameter of the laser output power variation versus time is likely the most widely used method for characterizing the width of well-behaved pulses i.e., those that can be described by a simple pulse-shape function; it should be noted that this is relatively insensitive to any weak background noises, as is frequently the case with laser pulses. There are four stages involved in producing these powerful nanosecond-long pulses by Q-switching (Loiko et al., 2016). First, the cavity losses are adjusted to a high level to prevent the laser transition from amplifying. A high multiple of the saturation energy can be reached by the stored energy, which is primarily constrained by the spontaneous emission. The energy channeled into the laser system by the pumping mechanism builds up in the upper laser level as population (inversion). The losses are abruptly lowered to sufficiently low amounts to remove any further obstructions to the laser amplification process after the maximum energy accumulation has been achieved. The gain saturates as a result of the temporal accumulation of intracavity power reaching the gain medium saturation energy. The output power now decays as the huge intracavity power quickly uses up the remaining stored energy; the laser pulse stops when the population in the upper laser level drops below the threshold value. It simply takes a few nanoseconds to complete. To avoid the possibility that any residual pump energy could repopulate the upper laser level past the threshold and cause secondary, laser action to begin, the cavity losses are raised once more after the laser pulse ends. Peak powers are typically larger than what would be possible from the same gain medium under CW operating conditions, and laser pulse energies in the region of a few millijoules are easy to obtain.

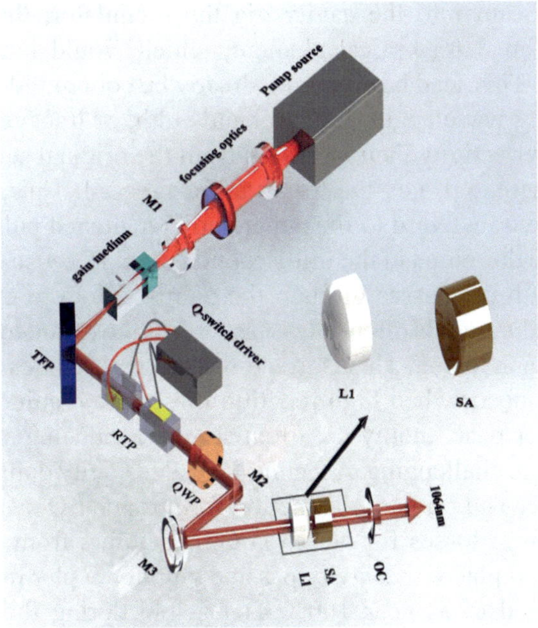

Figure 4.8. Q-switch.

Source: Yang, H., Li, Y., Wang, W., Feng, J., & Zhang, K. (2023). Simultaneous manipulation of the temporal and spatial behaviors of nanosecond laser based on hybrid Q-switching. *Photonics, 10*(2), Article 2. https://doi.org/10.3390/photonics10020227, Creative Commons License.

There are essentially two approaches to achieve the switching of the resonator losses between high and low: active Q-switching and passive Q-switching. Externally controlled variable attenuators are made up of active Q-switches. Although mechanical devices have been employed for this purpose, AO or EO devices are more frequently used; these devices operate similarly to the AO/EO modulators discussed above. Shortly after the electrical trigger signal arrives, an energetic laser pulse forms; the energy stored in the gain medium has a significant impact on the pulse energy and duration. The coupled rate equation pair can be used to model the actual evolution of the laser pulse and the related entities of gain and level populations. However, in this model, the coupling coefficient between photons and active emitters in the upper laser level is also a function of time, connected to the temporal characteristics of the Q-switch(Figure 4.8). It should be mentioned that using active Q-switches provides a number of benefits over passive ones.

Specifically, some Q-switch designs enable the coupling of an externally generated laser beam into the cavity via the modulator; this eliminates the requirement for an extra optical element, which would inevitably result in additional losses. This seed beam might already be configured for a few desired features, including wavelength, limited bandwidth, or transverse mode profile (TEM_{00}). The laser activity then increases from the original seed photons when the Q-value is increased, i.e., the losses are decreased. Thus, the traits of the seed radiation are transferred to the generated Q-switched pulses. Q-switching can achieve pulse durations in the millisecond range. Because the amplification process needs multiple passes through the gain medium to exhaust its stored energy, the pulse duration is often at least as long as the resonator round-trip time and frequently much longer. This is true if the resonator losses and/or laser gain are low; this can occur when high repetition rates are required, in which case significantly lower peak energy are often stored. Producing short nanosecond pulses may become challenging. A technique called cavity dumping can be used to solve this issue. The primary distinction from regular Q-switching is that an active mirror, whose losses for output coupling differs from full reflection to full transmission, replaces the cavity passive output coupler mirror. The mirror reflectivity is tuned to as near 100% as feasible during the Q-switch pulse generating phase; this effectively creates a closed (very low loss) resonator (Ershkov et al., 2015). Regardless of the time needed for pulse buildup, the energy from the circulating laser pulse is dumped out of the cavity within one resonator round-trip time once all of the energy stored in the gain medium has been transferred into it. The saturable absorbers, which are materials whose transmission rises as the light intensity beyond a specific threshold, are the foundation of passive Q-switching devices. The widely used materials for this include passive semiconductor devices, ion-doped crystals like Cr: YAG. Since the photons coming from the active medium are mostly absorbed, the cavity losses are initially significant. To ensure that there is still some laser action to occur and that a substantial quantity of energy is subsequently stored in the gain medium, the absorptivity is set somewhat high. With this kind of seeding, the laser intensity progressively rises until it reaches a point where the absorber is saturated. This quickly lowers the resonator loss, which causes the laser power to increase more quickly. Eventually, all of the energy is taken out of the gain medium, and the pulse stops. Following the pulse, the pattern repeats as the saturable absorber returns to its high-loss state. Regarding active Q-switching, the coupled rate equation pair can be used to model the temporal evolution of the laser pulse, laser gain, and level populations. Passive Q-switching is easy to use and economical; however, the pulse energies are usually lower, and external triggering of the pulses is not possible, which is detrimental in time-sensitive experiments that call for mutual synchronization. If ultrashort pulses from the picosecond to femtosecond regime are needed, Q-switching is no longer appropriate; instead, a technique called mode locking must be used. The method necessitates a broadband gain medium and a large number of longitudinal modes that resonate in the laser cavity above threshold within the gain bandwidth. The laser light builds up to circulating, short pulses if a fixed-phase relationship between those many oscillating modes is then introduced for brief periods of time. The factors affecting the pulse behavior for individual, full round trips remain same in the steady-state limit. The resonator round-trip time (τ_{CRT}), which is usually a few nanoseconds and corresponds to pulse repetition rates of up to 100 MHz, is equal to the repetition period between the emitted pulses if it is assumed that the resonator dimensions are such that only a single pulse circulates. Individual pulses have a much shorter duration, usually ranging from a few femtoseconds to a few picoseconds, largely determined by the number of locked modes. The peak strength of a single pulse can be higher than the average power in the pulse train, which is related to this difference in time scale. A gas laser-active medium bandwidth only allows a small number of modes, whereas a Ti: sapphire laser ultrawide bandwidth allows for thousands of modes(Goodfellow, 1986). The laser emits on a collection of separate laser modes at different frequencies, and these modes fluctuate independently of one another. A different behavior of the laser output is seen if all modes function with a fixed-phase relation among themselves. The laser modes with the specified periodicity will constructively interfere, producing an intense, short light pulse, as opposed to a comparatively random or continuous output intensity. The laser output pulses occur separated in time by $\tau_{CRT} = 2L/c$, or the time it takes for the light to accomplish exactly one round trip in the laser cavity, if only one such pulse periodicity is encountered in the cavity at any given time. The number of in-phase oscillating modes is correlated with the duration of each mode-locked laser pulse, or $\delta\tau_{MLP}$. The wider the bandwidth, the shorter the random variation of the laser pulse. In lasers with a small number of oscillating modes, the interference between the individual modes causes a beating effect in the laser output, which shows up as intensity fluctuations; in lasers oscillating on thousands of modes, these interference effects tend to average out and the output laser intensity is nearly constant. As with pulse-shortening by Q-switching, certain techniques to accomplish mode locking in a laser can be categorized as active or passive. In active mode locking, either the round-trip phase is periodically changed or the resonator losses are periodically modulated. Regarding Q-switching, the majority of these active devices are built on top of AOMs or EOMs(Chen et al., 2023). Ultrashort pulses of picosecond duration are produced as long as the modulation is synchronized with the resonator round-trip time (τ CRT). It should be mentioned that the modulator and cavity effects interact to determine the actual pulse duration; specifically, a gain bandwidth negatively determines how short the pulse can get. Mode locking can be implemented in three different ways: synchronous mode locking, AM mode locking, and FM mode locking. FM mode locking is the widely used method of active mode locking in the frequency domain. A sinusoidal amplitude modulation of the laser light oscillating in the cavity is produced by an AOM in the laser cavity that is driven by an appropriate electrical signal wave.

AM mode locking results from the modulation process, which may also be viewed in the time domain. The AOM in this instance functions as a shutter to the laser light wave that circulates within the cavity; to do this, it is powered by an electrical signal wave that sets the modulator device to various "open/ON" and "closed/OFF" states, altering the transmitted wave amplitude. Only one laser light pulse will pass through the cavity if the modulation frequency (f_{mod}) and the cavity round-trip time (τ_{CRT}) are synchronized. Because the same portions of the traveling wave pattern are continually attenuated as it passes through the cavity, it is sufficient for the actual modulation depth, or the amplitude difference between the open and closed states, to be of the order of 1% only. Several physical impacts must be considered while building true mode-locked laser systems. Phase dispersion in the cavity and nonlinear optical effects are the significant factors. The effect is linked to the fact that waves with varying frequencies move through optical medium at varying speeds, resulting in a distinct phase shift after passage. The phases of all oscillating modes cannot be locked any longer if the related net group velocity dispersion (GDD) over the entire cavity is too big.

This reduces the participating bandwidth, which in turn lengthens the mode-locked pulses. In order to compensate for the intrinsic positive GDD of laser cavities with conventional optical components, this fault can be addressed by adding optical elements to the cavity that display negative (anomalous) net GDD. There are two primary groups of techniques for negative-GDD shortening of optical pulses. By passing the pulses through an optical element with an appropriate degree of chromatic dispersion, such as chirped mirrors, grating pairs, prism pairs, or lengths of optical fiber, the chirp is eliminated in linear pulse compression.

4.10. COMPRESSION

The process of nonlinear pulse compression is two-step process. To create a chirped pulse with a longer duration than the initial pulse, the optical bandwidth is first extended, usually by a nonlinear interaction technique (self-phase modulation). Strong linear (dispersive) compression comes next, eliminating both the original and additional chirps. In order to lower the temporal peak strength of the laser pulse to below the damage threshold within the amplifier gain medium, nonlinear pulse compression is frequently used in the power amplification of ultrashort pulses. In this case, the amplifier stage is positioned between the stretching and compressing stages. Prism or grating pairs serve as the foundation for the most pulse compressors and stretchers, each of which has benefits and drawbacks. Although using gratings to make large negative or positive GDD is simple, their disadvantage is that they have very high losses (diffraction and absorption), making them inappropriate for intracavity applications.

The majority of laser systems are polarized, which typically indicates linear polarization, where the wave oscillation plane electric field is perpendicular to its propagation direction and remains stable throughout time. This is expected given that a standing wave must form in the cavity in order to be amplified in the gain medium and produce laser light output. The linear wave polarization must exist. However, because the initial spontaneous photons that self-start laser activity are random, the laser wave would begin to oscillate in an arbitrary polarization plane after switch-on if the gain medium and the laser cavity were both homogenous and isotropic. However, due to transient instability in the cavity, initial polarization state is typically unstable and may randomly shift between different directions. However, if there are negligible variations in the gain/loss differential for different polarization orientations, these are frequently enough to provide linear polarization that is stable. Directional anisotropy and linearly polarized laser emission can be caused by a variety of events in the gain medium or laser cavity(Panajotov and Prati, 2013). These consist of either the resonator losses are polarization dependent, or the gain medium itself is, as is the case with many laser crystals. For the latter, a preferred linear polarization direction may be produced by even a small tilt in the optical component, sometimes just as a result of poor alignment. However, in most cases, an optical element polarization dependency is intrinsic. Examples of this include Brewster-angled optical surfaces that show almost low losses for the corresponding ρ -polarized direction, or wavelength-dependent devices like gratings, prisms, tuning etalons, etc. When the angle of incidence is the Brewster angle $\theta B = \tan^{-1}(n_2/n_1)$, where n_1 is the refractive index of the incident optical medium (air) and n_2 is for the other optical medium, light with the ρ -polarization, light that oscillates in the same plane as the incident ray and the surface normal, will not be reflected, according to the Fresnel equations.

4.11. POLARIZATION

For many applications, the 1% polarization purity is adequate, but for others, significantly greater polarization levels are needed. Changes in polarization direction of fewer than 10^{-4} are common in Raman depolarization measurements, for instance, and even higher values are needed in precision ellipsometry. Many optical components are used to improve the degree of linear polarization in order to obtain such purities. These comprise the most widely used absorptive (or dichroic) polarizers; thin film polarizers based on dielectric coatings; linear film polarizers for nanoparticles; and polarizing beam splitters based on birefringence, such as the Glan polarizers. By altering the angle between the initial laser beam polarization, linear polarizers can also be used for controlled, external laser beam attenuation. The internal orientation of the polarization-determining optical element within the laser cavity determines the polarization orientation of a laser beam that emerges from a source. The polarization is either vertical or horizontal to the laser unit base for convenience. A so-called half-wave plate or half-wave retarder can be used to correct the polarization plane if it is incorrect for the specific optics. The optical propagation axis is parallel to the refractive sides of a double-refracting uniaxial crystal plate that makes up these. Coherence is one of the inherent characteristics of laser radiation; it is linked to the stimulated emission amplification process, which establishes the phase relationship between the photons from various emitters in the active medium. One can thus distinguish between two distinct characteristics of coherence, namely spatial and temporal coherence, based on the amount of space and duration of this fixed-phase relation. The associated coherence feature is also imprinted on the laser output beam.

CHAPTER 5

NMR SPECTROSCOPY

5.1. INTRODUCTION

Nuclear magnetic resonance (NMR) (Figure 5.1) can be explained as a radiofrequency irradiation that excites magnetically active nuclei in an external magnetic field, and the energy absorbed as a result of the irradiation (resonance) is recorded as an NMR signal. Different surroundings or environments for nuclei will result in different energies and, so, distinct signals; the sum of these signals creates a one-dimensional NMR spectrum.

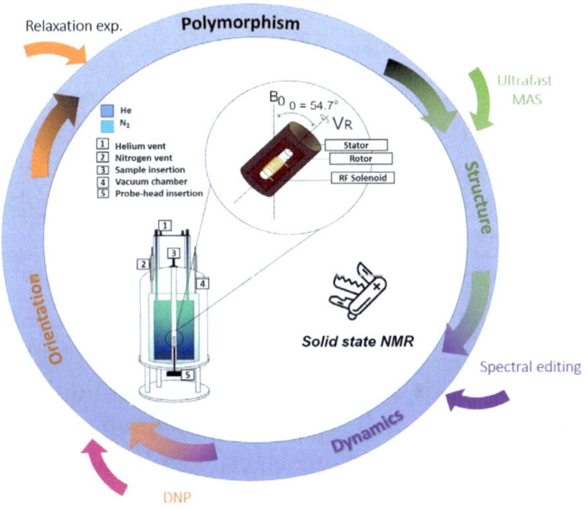

Figure 5.1. NMR.

Source: El Hariri El Nokab, M., &Sebakhy, K. O. (2021). Solid state NMR spectroscopy a valuable technique for structural insights of advanced thin film materials:

a review. *Nanomaterials, 11*(6), Article 6. https://doi.org/10.3390/nano11061494, Creative Commons License.

Line intensity plots versus resonance frequency are typically used to record NMR spectra (Figure 5.2), with the number of spectral lines indicating the number of non-equivalent nuclei determined by molecule structure and symmetry and reliant on spectral resolution (Wagstaff et al., 2013).

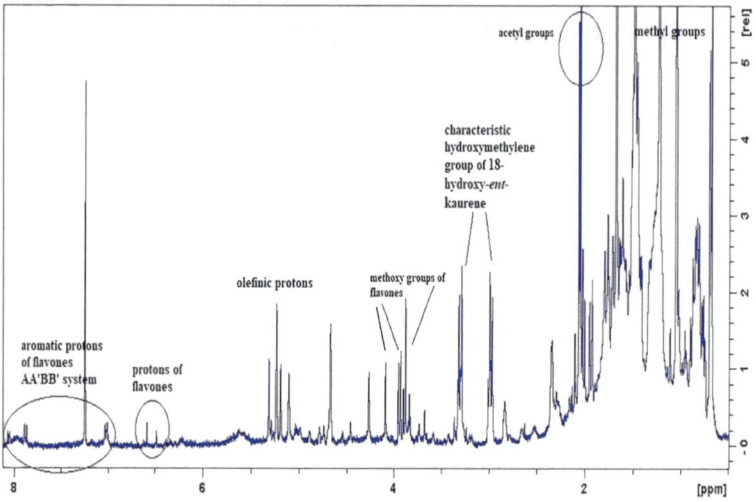

Figure 5.2. NMR spectra.

Source: Tomou, E.-M., Chatziathanasiadou, M. V., Chatzopoulou, P., Tzakos, A. G., &Skaltsa, H. (2020). NMR-based chemical profiling, isolation and evaluation of the cytotoxic potential of the diterpenoid siderol from cultivated Sideritiseuboea-Heldr. *Molecules, 25*(10), Article 10. https://doi.org/10.3390/molecules25102382, Creative Commons License.

One-dimensional NMR spectra of solids are with the resonance line orientations described by chemical shifts expressed in parts per million, or ppm. Splitting can be seen in resonance lines in spectra as a result of quadrupolar and dipolar coupling in solids or spin-spin coupling in liquids and solids. The environments of the nuclei affect the splitting. The number of nuclei in a sample determines the intensities of NMR signals, especially in solutions, which pave the way for structural investigation. It should be highlighted that full nuclear relaxation should be taken into consideration while recording the NMR spectra. Molecular structures in solutions have been explained by observing spin-spin coupling and precisely measuring the spin-spin coupling constants. There are at least 20 ^{19}F resonance lines in the solid complex $Fe(II)_5(bptz)_4(SbF_6)_{10}(CH_3CN)_{10}$ ^{19}F MAS NMR spectra (Chifotides et al., 2013). Since solids cannot detect the isotropic shift (^{19}F-$^{121,123}Sb$), the presence of three non-equivalent ions $[SbF_6]^-$ explains the number of resonances. Fast isotropic reorientations of $[SbF_6]^-$ ions in the solid-state result in six ^{19}F nuclei in $[SbF_6]^-$ being equal since the J(^{19}F–^{19}F) spin-spin coupling constant is not noticeable.

5.2. MAGNETIC FIELD

Understanding how nuclei react in the external magnetic field, why a single nucleus (^1H or ^{13}C) produces multiple lines in NMR spectra, and why these lines undergo additional splitting are all necessary for even these basic examples, which demonstrate the reasoning behind interpreting NMR data. Fundamental characteristics of nuclear systems exposed to an external magnetic field are chemical shift, spin–spin coupling, and quadrupolar and dipolar coupling (B0). Both quantum mechanics and semiclassical physics, which can both explain NMR, can be used to interpret the behavior of the nuclei when B0 is applied.

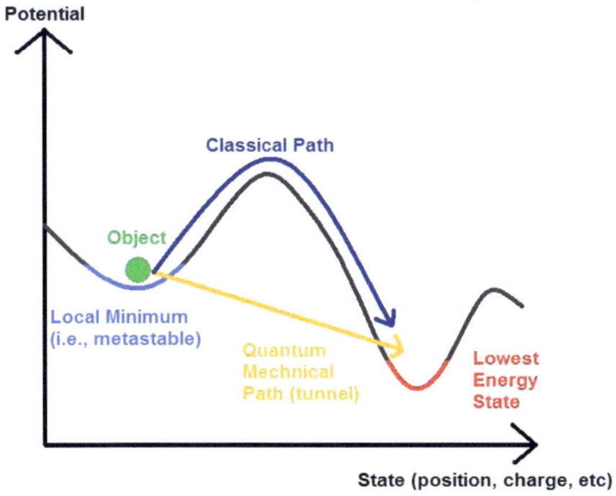

Figure 5.3. Quantum mechanics.

Source:https://en.wikipedia.org/wiki/Quantum_mechanics#/media/File:QuantumTunnel.jpg, Creative Commons License.

According to a quantum mechanics, (Figure 5.3)nuclear spin, a basic characteristic connected to a nucleus rotation around its axis, plays a key role. According to quantum mechanics, any system with distinct energy levels can be represented by an eigenvalue equation and specified by defined sets of quantum numbers. Similar to electron wave functions, nuclear spins in this context can be represented by wave functions ψ. Net magnetization is vital in the semiclassical explanation of NMR. When a sample is exposed to external magnetic field

B0, this magnetization is produced in it. If a nucleus has a non-zero angular momentum (P), it is deemed magnetically active and visible in NMR tests. If the number of protons and neutrons in a nucleus is equal, it will have zero spin; for instance, nuclei that are ^{12}C or ^{16}O are not magnetically active. However, it should be noted that, the majority of the elements in the Periodic Table have magnetically active isotopes, although their natural abundance varies. When external magnetic field B0 is present, nuclear momentum μ, which is classically equivalent to a tiny magnet, will respond. The response is highly dependent on the values of γ. When the same magnetic field is applied, this diapason then correlates to a wide range of resonance frequencies. The γ constant is positive for $_1H$, $_2H$, and $_{31}P$ nuclei and negative for ^{15}N, ^{29}Si, ^{17}O, ^{109}Ag, $^{117,199}Sn$, and ^{129}Xe nuclei(Bakhmutov, 2015).

The sign of γ is generally irrelevant in simple NMR experiments, but it is crucial in indirect nuclei detection based on heteronuclear multiple quantum coherence or HMQC, heteronuclear multiple bond correlation or HMBC, heteronuclear single quantum coherence or HSQC,used in solutions, and cross-polarization or CP in solids. The sensitivity of NMR tests conducted in the same external magnetic field is important. It is primarily dependent on the magnetogyric constant γ and the natural abundance of nuclei, or the nuclear nature. NMR sensitivity for ^{117}Sn, ^{13}C, and ^2H nuclei drop by a factor of 10^3, 10^4, and 10^6, respectively, in comparison to ^1H nuclei with complete natural abundance as a result of the influence of both of these features(Martins et al., 2000). This situation makes it difficult to detect these nuclei in both the solid and solution states. This is especially crucial for solids because of the strong dipolar internuclear interactions that can result in very broad resonances. Long-term accumulations are necessary for such solid-state NMR investigations in order to produce spectra with good signal-to-noise ratios. This is particularly difficult during extended periods of relaxation.

The spin numbers are also highly significant. Both in the solid state and in liquids, nuclei having a spin of ½ have considerably simpler NMR spectra. Since ^1H and ^{19}F or ^{31}P nuclei are completely naturally abundant and have $I = ½$, it is clear that they make the convenient targets for NMR studies, especially in solutions, where they exhibit strong dipolar interactions(Zheng et al., 2017). These interactions make it more difficult to identify them, inhibiting high spectrum resolution and necessitating the use of specialized NMR data collection methods. In quantum physics, the P and μ magnitudes are quantized. When the irradiation frequency v becomes close to the Larmor frequency of v_0, an NMR signal can be seen. Irradiation with frequencies ranging from 4 to 100 MHz is necessary for the detection of various nuclei put in a magnetic field (B0) of 2.3488 T(Gupta, 2013). This corresponds to the radiofrequency diapason. Technically, a powerful radiofrequency (RF) pulse can be applied to detect the macroscopic magnetization that results from the presence of magnetic field B0, or CW irradiation can be applied earlier. Initial energy absorption and subsequent emission occur when an NMR signal is observed. It is crucial that the NMR signal cannot be registered without this energy emission.

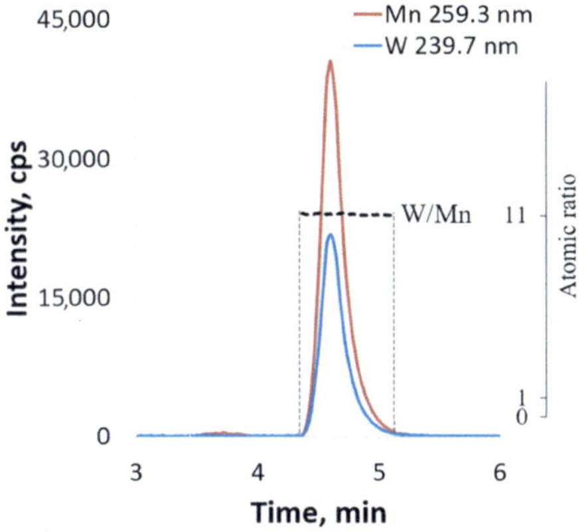

Figure 5.4. Nuclear relaxation.

Source: Korenev, V. S., Burilova, E. A., Volchek, V. V., Benassi, E., Amirov, R. R., Sokolov, M. N., & Abramov, P. A. (2023). NMR-relaxometric investigation of Mn(II)-doped polyoxometalates in aqueous solutions. *International Journal of Molecular Sciences, 24*(8), Article 8. https://doi.org/10.3390/ijms24087308, Creative Commons License.

5.3. ENERGY INTERACTIONS

Only an energy exchange between nuclear spins and their surroundings can lower the energy of an excited nuclear system. Nuclear relaxation, (Figure 5.4) which restores the original equilibrium state, is directly linked to this exchange. No NMR signal is seen when nuclear relaxation is absent or when it is present for a long time. Applying a strong radiofrequency pulse at the acting frequency ω_{rf} is known as pulsing irradiation. An oscillating and time-dependent RF field is created by this RF pulse, which instantaneously excites a wide frequency range and contributes to the external magnetic field B0. In liquids and solutions, where molecular movements are extremely rapid and high amplitude, Lorentzian resonance forms are typically expected. It should be mentioned that materials with the high-temperature phases or plastic crystals might experience such rapid isotropic motions.

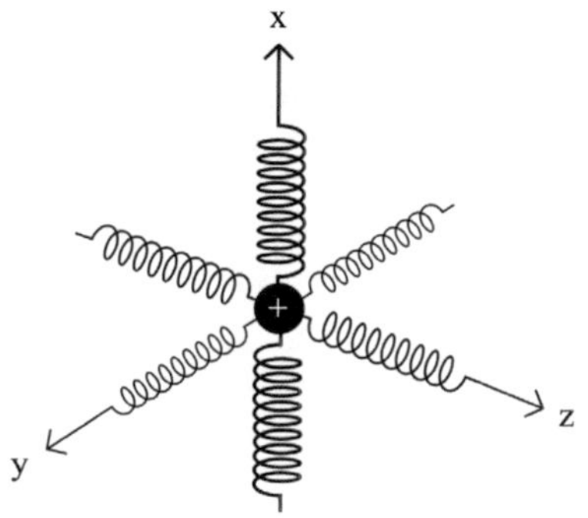

Figure 5.5. Lorentzian resonance.

Source: https://upload.wikimedia.org/wikipedia/commons/b/b9/Atom-spring.svg, Creative Commons License.

Lorentzian resonance (Figure 5.5)forms can also be seen in these solids. Gaussian curves or a combination of Gaussian and Lorentzian curves are typically used to mimic NMR resonances in hard solids(Taylor et al., 2002). Several significant findings can be drawn since the NMR line shape is crucial to researchers from both a theoretical and practical viewpoint. First, understanding the line shape aids in achieving high spectral resolution through the adjustment of shimming coils in solid and solution experiments. All of the detected resonances can be impacted by improper shimming techniques that produce an inaccurate shape. Second, two neighboring Lorentzian lines may overlap while two Gaussian lines may stay separate because the wings of the Lorentzian curves are noticeably longer than those of the Gaussian curves. When it comes to signal integration, this effect is crucial. Third, the theory of line shapes makes it possible to quantitatively describe spin states undergoing exchanges on the NMR time scale by calculating the signals within the context of a line-shape analysis. Dipolar and/or quadrupolar coupling calculations, which are crucial for solids, can also be performed using the line-shape analysis. However, the power NMR spectra, which are constructed using typical software techniques using the real and imaginary components, cannot be compared to the concept of the line shape. When the standard NMR spectra exhibit low signal-to-noise ratios or necessitate significant phase and baseline corrections, these power spectra may be helpful. The basic schematics include a radiofrequency reference generator, a transmitter, an amplifier, a receiver, a pulse generator, a powerful computer, a magnet, and an NMR probe, can be used to illustrate the principal schemes of the wide range of techniques currently used for NMR research.

An RF coil in an NMR probe trace the macroscopic magnetism in a material. When the coil in the NMR probe is precisely tuned to carrier frequency ω_0, maximal A values can be achieved since this signal has the form A $\exp(i(\omega_0 + \omega)t)$, where A is the signal amplitude, ω_0 is the carrier frequency, and ω is the required frequency offset. It is important to keep in mind that improper tuning might significantly alter the final NMR spectra. The nature of the solids and liquids actually has a significant impact on the sample magnetic susceptibility. The quality of each unit has a substantial impact on the signal-to-noise ratios in the resulting NMR spectra, even for a well-tuned NMR probe. The strength of the external magnetic field determines the NMR sensitivity when these devices are operating at optimal levels. Due to the extremely weak signal in the RF coil, it needs to be amplified, filtered, and combined with an intermediate frequency. The next step is to separate the signal components into real and imaginary sections. Following this separation, a digital signal is created using an analog-to-digital converter. A computer then processes the signal to produce the final NMR spectrum.

5.4. NMR SPECTROMETERS

Every step mentioned above is standard for every NMR process. However, solid-state and solution NMR spectrometers have different requirements.

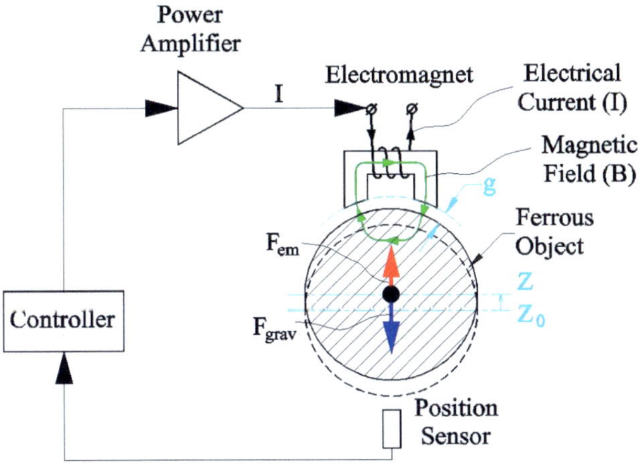

Figure 5.6. Permanent iron electromagnets.

Source: Filatov, A., Hawkins, L., & McMullen, P. (2016). Homopolar permanent-magnet-biased actuators and their application in rotational active magnetic bearing systems. *Actuators, 5*(4), Article 4. https://doi.org/10.3390/act5040026, Creative Commons License.

First, a solid-state NMR spectrometer should have a 12-bit analog-to-digital converter that can digitize spectral windows of up to 5 MHz, as opposed to a 16-bit converter for solution NMR, since solids would be characterized by NMR using spectral windows of 1 MHz or greater, unlike solutions. The latter allows for the observation of a 125 kHz maximum spectral window. The solid-state NMR spectrometers should have high-speed digitizers with significantly quicker speeds since the solid-state NMR spectrum spans a very wide frequency range. The uniform magnetic field needed for initial nuclear polarization and, consequently, for the measurement of high-resolution NMR spectra comes from a magnet.

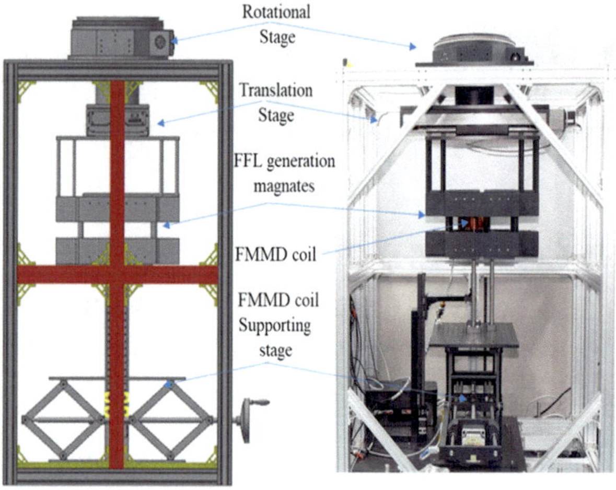

Figure 5.7. Wide-bore magnets.

Source: Jeong, J. C., Kim, T. Y., Cho, H. S., Seo, B. S., Krause, H. J., & Hong, H. B. (2024). MPI system with bore sizes of 75 mm and 100 mm using permanent magnets and FMMD technique. *Sensors, 24*(12), Article 12. https://doi.org/10.3390/s24123776, Creative Commons License.

Permanent iron electromagnets (Figure 5.6)that produced magnetic fields of 1 to 2 T, or a ^{1}H frequency of 100 MHz, were used in the first NMR investigations(Moresi and Magin, 2003). But the iron magnets are highly temperature-sensitive and comparatively unstable. Superconducting electromagnets, on the other hand, exhibit stability and good uniformity. The magnetic fields that have been developed may now operate at proton frequencies of up to 900 MHz, where field locking can increase the magnetic field stability, which is essential for high-resolution NMR investigations in solids and liquids. However, it should be mentioned that wide-bore magnets, (Figure 5.7)which can allow a variety of sample manipulations, are typically needed for solid-state NMR investigations.

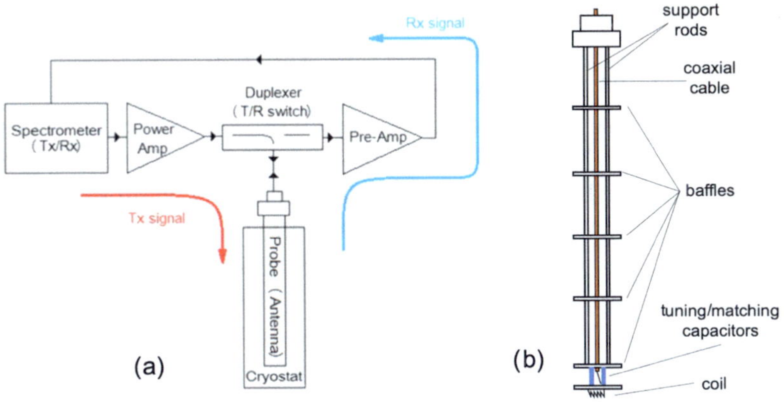

Figure 5.8. Wide-line NMR probes (a and b).

Source: Kolar, P., Grbić, M. S., &Hrabar, S. (2019). Sensitivity enhancement of NMR SPECTROSCOPY receiving chain used in condensed matter physics. *Sensors, 19*(14), Article 14. https://doi.org/10.3390/s19143064, Creative Commons License.

Since sensitivity and chemical shift dispersion, which are measured in hertz, rise proportionately to field strength, the long-term trend in NMR research has been predicated on the notion that only larger magnetic fields can result in better spectrometers. A sample needs to be positioned inside the stationary magnet that generates the powerful magnetic field. However, small magnets have been made possible by recent advancements in the manufacturing of compact, low-cost NMR magnets. Low-field NMR spectroscopy is a new field within NMR investigations, and the realization of small NMR magnets with field strengths up to 2 T has made it possible to create mobile NMR equipment. One crucial part of spectrometers is the NMR probe. In order to reduce inhomogeneity of the magnetic field, a sample is typically spinning in the NMR probe at a low rate (20 Hz). The rotation rate should be significantly greater (20 to 30 kHz) in order to decrease dipolar interactions in solids. The design of NMR probes varies significantly between solid and solution investigations. It depends on the requirements of the study; for instance, experiments using specifically made NMR probes can be double or triple resonance or gradient-field. Accurate temperature settings for the heating and cooling apparatus and the lack of visible temperature gradients through a sample are necessary for variable-temperature measurements in liquids and solids, which are typically between –120° and +120°C.

5.5. PROBES

Wide-line NMR probes (Figure 5.8)are typically used for solid-state experiments on static samples. The sample is positioned in a glass NMR tube that is perpendicular to the external magnetic field. Wide-line probes of this type can be fitted with coils, which are typically required for biomolecular research. The samples may typically be mechanically rotated by solid-state NMR probes at high spinning rates. When MAS NMR rotors are positioned at the magic angle with respect to external magnetic field B0, the strong dipole–dipole interactions in solids will be suppressed at high spinning rates. Zirconium oxide rotors of diameters of 7, 4, 2.5, and 1.3 mm can be used to regularly achieve spinning rates of 7, 18, or 25 kHz(Balinova et al., 2018). Finding several magnetically, chemically, or structurally non-equivalent nuclei in liquids and solids is one of the main tasks of NMR analysis. It is clear that the spectral resolution, which can be defined as the minimum distinct distance, expressed in hertz, between two peaks in an NMR spectrum is crucial. The line widths of NMR signals, expressed in hertz, in typical oxygen-free solutions define the spectral resolution in solution. The external magnetic field homogeneity determines the spectral resolution, and it can be enhanced using common shimming coil manipulations. In solutions, this process may consistently yield a spectral resolution of 0.1 to 0.2 Hz. Analog-to-digital converters transform the FIDs obtained by NMR into digital formats, which are then analyzed by computers. The size of the memory employed for this treatment determines the final resolution of NMR spectra following Fourier transformation. In high-resolution NMR, where the digital resolution is clearly a function of the sweep widths used for the data collection and the number of points utilized for the treatments, this size defines the digital resolution, which is very crucial. It should be highlighted that even careful adjustments using shimming coils that provide a high level of physical precision will be constrained by a limited number of points. The actual spectral resolution is determined by this value. Shimming coil adjustments are crucial for every sample since the predicted resonance lines are typically sharp ($\Delta v = 1/\pi T_2$) and the nuclear relaxation durations (T_2) are long in solutions. Gathering or collecting NMR data necessitates a large amount of computer capacity. However, because of short T_2 periods and strong dipolar internuclear interactions, resonance lines are typically broad in solids. Therefore, a suitable balance between sweep width and computer memory size is possible, and the shimming coil manipulations are not crucial in solid-state NMR. In order to enhance the resolution, to eliminate overlapped resonances, the number of points should be increased concurrently while the FIDs are subjected to a Gaussian function. When the strong external magnetic field is homogeneous and able to sustain the resonance conditions in every point throughout the volume of a sample, highly resolved NMR spectra can be produced. This is essential for liquid and solid NMR investigations. However, it is frequently not possible to satisfy the resonance criteria throughout an entire large heterogeneous sample. It is possible to homogenize the external magnetic field in a small volume while yet recording a high-resolution NMR signal from this small region. Repeating these procedures will enable the observation of target nuclei dispersed across the sample volume. This small volume can be spatially altered to record the subsequent signal. Typically, when a sample is placed in external magnetic field B0 for NMR observation, its nuclear spins are not separated. They have the ability to communicate with their surroundings.

5.6. COUPLING

These interactions may involve paired or unpaired electrons as well as the same or distinct kinds of nuclei. The phenomena of chemical shift, spin–spin coupling (homonuclear and heteronuclear), quadrupolar coupling, and dipolar coupling are caused by interactions with the nearest environments via chemical bonds. Since chemical shifts and internuclear couplings are the primary factors influencing NMR spectra, along with the number of resonances and their line shapes, researchers are always interested in them, regardless of the objects and their nature, a solution or a solid. Resonance line placements in NMR spectra can be used to define chemical shifts δ, which often indicate isotropic chemical shifts. Two $^{13}C\{^{1}H\}$ resonances that belong to group CH_2 and two equivalent groups CH_3 are displaced on the frequency scale by the carbons in the $CH_3CH_2CH_3$ molecule. A local magnetic field acting on the detected nucleus is thecause of the displacement. An extra magnetic field that protects the external magnetic field is produced when electrons around the nuclei move in external magnetic field B0. Tetramethylsilane or TMS for ^{1}H and ^{13}C nuclei, an aqueous solution of H_3PO_4 for ^{31}P nuclei, or a solution of CF_3COOH for ^{19}F nuclei are a few examples of other substances that can be used as standards (Ramirez et al., 2002). It should be highlighted that an internal reference is clearly preferred because of the significant influence of magnetic susceptibility, which varies in various samples. Any resonance with a known chemical shift that is referred to as a standard can be utilized, or a standard can be introduced into a sample. It is important to note that digital resolution, which is defined in ppm or hertz as a distance between two points, is necessary for accurate assessments of chemical shifts, even at very good spectral resolution. Chemical shifts in NMR spectra in solids and liquids are important factors used in structural analysis, where the connection between chemical shifts and molecular electronic distributions is crucial. Therefore, assuming the other parameters remain constant or fluctuate negligibly in a group of comparable compounds, chemical shifts and charges on atoms or electronic density can be connected (linearly). In these compounds, the average energy of bond order Q and electron excitation E may stay similar. This relationship serves as an example of the principle that researchers frequently use when analyzing NMR spectra and signal assignments: A diminishing electronic density on an atom is represented by the low-field displacement of an NMR signal. For aromatic compounds, when the $\Delta\rho$ charges are determined via quantum chemical calculations, the linear correlations $\delta(^{1}H)$ vs. $\Delta\rho$ are well established. In the same framework, the Hammett/Taft constants of groups in pyrazines, 2-substituted pyridines, and benzenes are expected to correlate with proton chemical shifts. For different groups of organic compounds, linear correlation between $\delta(^{13}C)$ and atomic charge have also been demonstrated. One of them, derived for solutions of polyfluorinated olefins and allyl cations. It is evident that the total MNDO atomic charges have a linear correlation with the 21 $\delta(^{13}C)$ values observed for sp^2-hybridized carbon atoms(Lodewyk et al., 2012). The linearity is not improved by adding diamagnetic adjustments. These correlations by themselves are not suitable for quantitative assessments of atomic charges since the values of atomic charges are dependent on the technique utilized in their quantum chemical calculations.

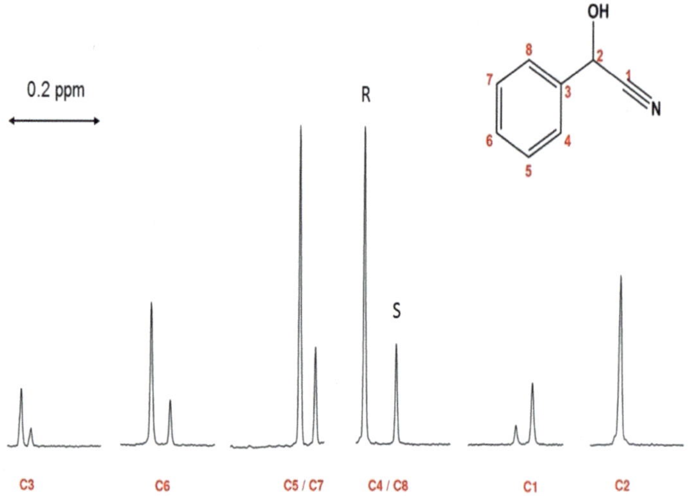

Figure 5.9. ^{13}C NMR spectra.

Source: Lankhorst, P. P., Van Rijn, J. H. J., &Duchateau, A. L. L. (2018). One-dimensional 13C NMR Is a simple and highly quantitative method for enantiodiscrimination. *Molecules, 23*(7), Article 7. https://doi.org/10.3390/molecules23071785, Creative Commons License.

The signals in ^{1}H and ^{13}C NMR spectra(Figure 5.9) can be assigned with the aid of the observed tendencies. They pave the way for computer analysis of ^{13}C NMR spectra, in which these tendencies are used to assign ^{13}C signals. The issue becomes complex and uncertain when heavy nuclei other than ^{1}H and ^{13}C are involved. Changes in the isotropic chemical shifts recorded in solutions are implied by the dependence of $\delta(^{13}C)$ on π charges (q_π) in aromatic compounds. It is interesting to observe how the charge q_π influences three-dimensional ^{13}C changes in materials. The first crucial stage in any structural investigation based on one-dimensional NMR (solids or solutions) is the comprehensive assignment of signals in the NMR spectra. In general, it is possible to forecast the chemical shifts for signals in a novel compound NMR spectra. There are three methods for assigning NMR signals and predicting δ values. They are database searches, quantum chemical calculations, and calculations based on empirical systems. The widely used method is likely a database search based on a structural analogy between known and unknown compounds. This is because the search makes use of the huge experimental data that establishes spectrum/structure relationships and can show the δ values as a function of atomic coordination numbers. They can show atomic oxidative states, H bonding, etc. Even with heavy nuclei, this method can be rather effective. Chemical shift values for existing or hypothetical structures can be computed using the quantum chemical calculations and the quantum mechanical theory of NMR. The experimental NMR spectra are then rationalized by these computations. The prediction of chemical shift values is confirmed by the fact that many of the computations are highly successful or exhibit strong linear correlations between the theoretical and experimental values. The $\delta(^{1}H)$ values for hydride ligands in transition metal complexes are accurately predicted by fully relativistic (FR) density functional theory (DFT) calculations than by non-relativistic (NR) and scalar relativistic (SR) calculations. These computations are not simple; they are intricate. There are more simple methods; the chemical shifts of the ^{1}H, ^{13}C, ^{15}N, ^{17}O, and ^{19}F nuclei can be determined using the semi-empirical MNDO methodology. However, the prediction of ^{1}H and ^{13}C chemical shifts based on empirical systems is easier and more accurate when it comes to the ^{1}H and ^{13}C NMR spectra.

5.7. CHEMICAL SHIFTS

Apart from the diamagnetic, paramagnetic, and magnetic anisotropy components that contribute to the overall chemical shifts, the presence of distinct isotopes within a molecule also causes isotropic chemical shifts. The isotropic effect on the ^{1}H chemical shift is best understood by the ^{1}H NMR spectra of chloroform ($CHCl_3$). The ^{1}H nuclei that are adjacent to the non-magnetically active isotope

^{12}C, which has a natural abundance of 98.9%, are represented in the 1H NMR spectrum. A weaker doublet for ^{1}H adjacent with the isotope ^{13}C, which has a 1.1% natural abundance, also shows up in addition to this signal(Wang et al., 2013). The spin-spin coupling $^{1}H–^{13}C$ explains this doublet; however, the doublet center, which corresponds to its chemical shift, is detected at other higher field than the intense resonance. Therefore, the presence of the pairings $^{1}H-^{12}C$ and $^{1}H-^{13}C$ in $CHCl_3$ can account for the observed $\delta(^{1}H)$ discrepancy. At their relatively high-field displacement of 0.04 ppm, the solution NMR spectra of H_2 and HD molecules display a singlet and a triplet. Electrons in nuclear settings are not the source of the isotropic chemical changes. The behavior of isotopes is clearly different in intramolecular dynamics, which includes the vibration and rotation factors. It is acknowledged that primary and secondary isotope shifts can reveal the effects, which differ depending on the nucleus. For distinct isotopes in the same chemical (molecular) system, primary isotope shifts manifest and can be quantified in NMR spectra in solutions. Protons and deuterons in the same molecule will exhibit chemical shifts that differ by 0.1 ppm or less due to the primary isotope effect. NMR tests on nuclei with surroundings containing different isotopes reveal secondary isotope shifts. For heavy nuclei, such as ^{119}Sn nuclei, the secondary isotope effects can reach 6 to 9 ppm, whereas the impacts are moderate or large for ^{13}C, ^{15}N, and ^{31}P nuclei. By frequent spins stimulated by radiofrequency (RF) irradiation to their initial equilibrium state, nuclear relaxation makes it possible to see an NMR signal. The same process is represented by the frequency and time-dependent data collected by pulse NMR investigations. NMR applications are divided into two distinct groups: NMR relaxation and NMR spectrometry. Understanding the theory of nuclear relaxation is necessary to conduct even the basic NMR experiments and to interpret the findings. The former is most common among researchers who employ NMR for structural investigations. In order to prevent misinterpreting NMR spectra, it is also essential to comprehend how nuclear relaxation alters the line widths and resonance shapes in solids and liquids. By identifying and characterizing motions on a large frequency scale, relaxation studies can help us comprehend molecular dynamics. They can also facilitate structural diagnostics by determining interatomic distances and specific relaxation requirements.

5.8. NUCLEAR RELAXATION

Nuclear relaxation occurs through an energetic exchange between nuclear spins stimulated by radiofrequency energy and their surroundings. The exchange can only occur when fluctuating magnetic fields generated within the lattice are present. The nuclei and electrons are physically found in molecules, and these magnetic fields are produced by nuclear and/or electron magnetic moments.

The molecular movements (thermal) cause fluctuating magnetic fields, which in turn cause nuclear relaxation. From slow translational motions and rotational reorientations to extremely quick molecular vibrations with a frequency range of 10^{12} to 10^{14} Hz, the frequency scale of molecular motions can be large. When the fluctuating frequencies are around the Larmor resonance frequency, v_0 or ω_0, spin-lattice relaxation can be effective(Grant et al., 2002). The movements of molecules in solids and liquids differ greatly. Molecules may undergo slow and fast translational motions, rapid thermal rotational reorientations, or both. There are a variety of intramolecular motions, including segmental motions in polymeric structures and rotations around single or partially double chemical bonds. As a function of motional frequency, these motions contribute to nuclear relaxation in unique ways. The molecules can be spheres despite their varied motions, and a single correlation time can adequately explain the motions.

According to the Stokes-Einstein-Debye theory for liquids and solutions, the bulky viscosity of medium η (solvents) and molecule sizes determine the molecular motion correlation times. Molecular motions with high or low amplitudes can be classified as either isotropic or anisotropic. This classification is crucial because isotropic and anisotropic motions affect relaxation times in both solutions and solids, as well as NMR spectra and resonance line forms, especially in solids. A molecular motion is considered isotropic when it is described by a single τ_C value and a single E_a value. It is acknowledged that when experimental errors in calculating T_1, $T_{1\rho}$, and T_2 are kept to a minimum, quantitative or semiquantitative interpretations of relaxation durations in the context of structural studies or characterizations of molecular dynamics will be reliable.

The longitudinal and transverse components of the total nuclear magnetization recover to an equilibrium state with time constants T_1 and T_2. Because of the quick and anisotropic molecule reorientations, liquids typically exhibit basic exponential form of nuclear relaxation. The spin systems with more than two energy levels i.e., coupled relaxation can exhibit non-exponential behavior even in liquids. It should be mentioned that solids frequently exhibit non-exponential relaxation. The emergence of anomalous relaxation behavior in NMR studies necessitates a thorough comprehension of its nature and the cause of the non-exponentiality, in addition to expertise in data processing and T_1 or T_2 time computation. A number of factors can impact experimental measurements and lower the possible accuracy of relaxation time determinations. The temperature should be consistent within ± 0.5°C because relaxation experiments typically last long and T_1/T_2 times rely on temperature. This is especially crucial for solids that MAS NMR probes since the high spinning rates raise the temperature.

5.9. PULSES

Any of the several methods that are available should be used to precisely calibrate the 90° and 180° radiofrequency pulses. Given that temperature is known to affect pulse length, this calibration ought to be carried out for every temperature. Ineffectively non-exponential magnetization recovery curves are produced by poorly calibrated pulses, whereas exponential recovery curves are actually produced. T_1 and T_2 computations may contain significant inaccuracies if such curves are treated using the normal fitting procedure of NMR spectrometers. It is clear that the signal-to-noise ratio has an impact. Low signal-to-noise ratios make it harder to determine integral intensity accurately. In τ regions, where line intensities are near zero, this is especially crucial. Therefore, the τ values determine the errors in identifying intensities in the presence of spectral noise. Radiofrequency pulse inhomogeneity is a technical issue. Only at the center of the radiofrequency coil is the radiofrequency field homogeneous. The pulses grow imperfect as they move away from the center, which could cause significant departures from the behavior of exponential relaxation. It should be evident that the total relaxation delays used in pulse sequences match the nuclei relaxation throughout each measurement cycle. The signal intensities will be severely distorted when the delays are too short. It is important to keep in know that a delay near a T_1 value only recovers about 63% of the equilibrium magnetization, whereas a delay of $5T_1$ can accomplish a 99% recovery of the equilibrium magnetization(Stanisz et al., 2005). The experimental relaxation curves should be accurately described since the kind of nuclear relaxation is typically unknown before. To distinguish between exponential and non-exponential processes, the τ values employed in inversion-recovery studies might vary from $0.1T_1$ to $3T_1$. It should be mentioned that a good indicator of complex nuclear relaxation is the emergence of overshoots in the inversion–recovery curves, especially for τ values near T_1 and $2T_1$. Errors in T_1 and T_2 readings are typically less than 3 to 5% due to routine NMR modifications. The quality of the experimental results and the relaxation model employed determine how well an experiment is treated, which is a crucial step in interpreting relaxation data. Both exponential and non-exponential magnetization recovery are possible in solutions, as in the case of coupled relaxation. The number of relaxation components can determine whether the process becomes bi-exponential or multi-exponential. In solids, NMR relaxation is not simple or easy. The total recovery curves (echo decays) might significantly depart from an exponential function even if each site in inhomogeneous solids relaxes exponentially with its own relaxation period. Spin diffusion relaxation is an exception. Since every spin in a sample relaxes in the same way, this relaxation is often exponential. It is important to note that spin diffusion is useless for short τ delays, like those used in inversion-recovery investigations. Magnetization recovery is proportional to time ($\tau^{1/2}$) in such short τ areas(Zhou et al., 2006). Magnetization recovery then becomes exponential as τ values rise.

The extensively used method for handling relaxation curves is based on comparing experimentally measured signal intensities to a model function to find the fitted parameters. The relaxing time is one of these variables. In order to arrive at the optimal fitting, researchers utilize practical computing programs that are based on the Levenberg-Marquardt algorithm. Uncertainty may result especially if the model function being utilized for computations is complex. The model may really be a stretched exponential, a sum of the Gaussian and exponential functions, or a sum of two or more exponentials. Tests utilizing bi-exponential and stretched exponential models can yield similarly satisfactory results with certain combinations of the T_1(short) and T_1(long) components. The curves can be well-fitted to exponentials, bi-exponentials, or stretched exponentials when the short and long relaxation T_1 components are proportionate, indicating that the treatments are not entirely single valued. The experimental relaxation curves may appear exponential despite their more complicated nature, or they may be equally well handled by stretched or bi-exponentials. The character of the relaxation can be more clearly identified by representing the signal intensity as a function of the τ time in semilogarithmic coordinates. These coordinates should display straight lines for exponential processes. Since their emergence is unpredictable and difficult to explain, artifacts in NMR spectra are typically referred to as such. Interpreting relaxation experiments is made more difficult by artifact effects. The relaxation behavior of nuclei, especially protons, can also be impacted by the radiation damping effect, which is caused by the interaction of an RF coil with a sample bulk magnetization. The measured ^1H T_1 values will be less than the actual values for very short damping times, and the exponential relaxation itself may stop being exponential. Research using detuned NMR probes or isotope dilution to lower the $_1H$ content in a sample can significantly lessen the effect. Artificial attenuation of echo signals can be detected by measuring the spin–spin relaxation times, T_2, using echo pulse sequences. Experiments on samples with extremely short T_2 times for brief initial τ delays introduced in the pulse sequence show the attenuation. For low τ delays, the initial echo intensity rises as τ values climb and then often falls near an exponent. The vortex effect can cause artifact phenomena in solids spinning at high rates; however, it has no influence on NMR relaxation investigations on solutions because of the low spinning rates. The relaxation process may become non-exponential due to inaccurate relaxation time values caused by imperfect RF pulses. In MAS NMR investigations with 50% filled rotors, this undesirable behavior may appear. The RF field is in homogeneous, the pulses are again imprecise, and a powder actually takes on a solid-vortex structure as a result of spinning at high speeds. Even for other NMR investigations that need accurate RF pulse calibrations, this effect may be significant.

5.10. SPIN

In relaxation NMR investigations conducted on high-spinning samples that contain liquids or liquid components, such as experiments on suspensions of heterophase samples, another vortex mechanism may appear. A fast-relaxing component can be seen when spinning, as the initial exponential spin-lattice relaxation shown in the static samples can change to a bi-exponential process. Researchers are interested in molecular dynamics because it has a significant impact on the physical and chemical characteristics of a wide range of systems, including solid-state materials and simple and complex molecules in solutions. Depending on the frequency and nature of motions, molecules or chemical groups may have an impact on the NMR parameters that are observed. Many observable resonances, quadrupolar splits, dipolar couplings, their line widths and shapes, chemical shifts, spin–spin coupling constants, and relaxation periods in solutions and the solid state can all be altered by intramolecular and intermolecular reorientations. To ascertain the frequencies, and activation energies of molecule reorientations, NMR spectroscopy quantitative investigation of molecular mobility must therefore consider each of these variables. The kinds of molecular motions and how quickly they occur vary. Dynamically heterogeneous solids such as porous materials and glasses constitute a unique class. These systems can be made as solids having an open matrix filled with a liquid, such as biological tissues, plasticized polymers, gels, and zeolites. Depending on the matrix structural properties, the small liquid spaces can produce extremely complicated liquid dynamics. Translational diffusion is also interesting and needs quantitative characterizations, especially in immobilized catalysts with solvents present. Dynamic NMR (DNMR) studies focus on the quantitative measurement of these dynamics in solids and liquids(Reichert, 2005). The time scales of the motions under study determine the experimental procedures and DNMR approaches. Measurements of relaxation times are used to identify the correlation times (τ_C) that define the rapid molecular reorientations. Slower molecular motions linked to quadrupolar coupling tensors or reorientations of chemical shift anisotropy tensors take place on the same time scale as the free induction decay (FID) time, or faster. Line-shape analysis can be used to characterize these reorientations since they greatly alter line forms and conform to the condition τ_C T_2. Changes in orientation-dependent NMR frequencies allow for the detection of slower dynamics, which may then be investigated using exchange NMR spectroscopy. NMR spectra can be recorded over time to study very slow processes on the NMR time scale. Regular NMR can be used to quantitatively study the kinetics of a chemical reaction, which is regarded as a special type of chemical motion, when it occurs significantly more slowly than the collection of FID data and when the time required to mix reagents and transfer a sample into the magnet is minimal. This work can be regarded as both kinetic and spectroscopic in certain ways. Accurate measurements of time-dependent integral intensities in the NMR spectra that decrease or increase for initial compounds or products, respectively, are the goal of the spectroscopic component.

Conventional methods can attenuate the unwanted strong signal from the solvent (HOD) for substances with poor solubility. There are several types of NMR tubes available for NMR investigations in solutions, such as coaxial tubes, high-pressure NMR tubes, conventional tubes (different diameters), and tubes for microsamples to obtain the highest resolution. The center of the RF coil, where the radiofrequency field is uniform, is where the tubes for microsamples should be placed.

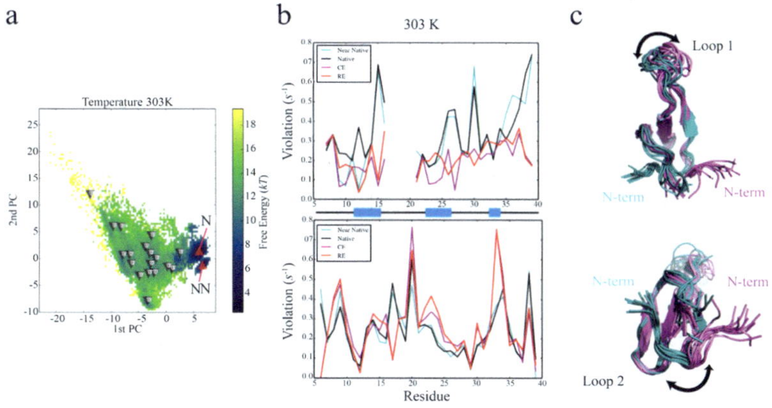

Figure 5.10. Nuclear Overhausereffect.

Source: Nichols, P. J., Born, A., Henen, M. A., Strotz, D., Orts, J., Olsson, S., Güntert, P., Chi, C. N., &Vögeli, B. (2017). The exact nuclear Overhauserenhancement: recent advances. *Molecules, 22*(7), Article 7. https://doi.org/10.3390/molecules22071176, Creative Commons License.

Commercially available NMR tubes that are made to cover samples can be used to produce air sensitive samples in a vacuum or in an argon or nitrogen environment. The solvents and solutions must be oxygen free to obtain the best resolution and quantify nuclear Overhauser effect (NOE) relaxation times(Gerig, 2008)(Figure 5.10). Deoxygenation can be accomplished using three or four cycles of freeze-pump-thaw. If required, the samples can then be sealed in an inert atmosphere or under vacuum. An easy yet crucial step in getting prepared for NMR investigations is adjusting the spectral resolution. The ability of an NMR spectrometer to discern lines in the NMR spectrum that are near to one another is sometimes referred to as spectral resolution. For resonance frequency v_0, it can be defined as $R = v_0/\Delta v$, where v is the lowest frequency difference that can be recognized. The resolution is directly related to the natural width of a Lorenz-shaped signal from uncoupled spins in an ideal magnetic field, which is determined by the spin–spin relaxation time T_2. The number of shimming coils is typically increased in the higher magnetic fields, and shimming manipulations can become increasingly complex. In these situations, mapping the shims to obtain good resolution can be greatly aided by the pulsed field gradient approach. To prevent confusion, it should be emphasized that the spectral resolution as measured by line width cannot be higher to that ascertained by a natural T_2 time, and computer hardware and software should be able to support the recording of the test NMR spectra in order to provide adequate digital resolution.

Since the quality of the final NMR spectra determines the reliability of NMR interpretations based on the determination of integral intensities and chemical shift values, they should be recorded with precise phasing and baseline correction methods following Fourier transformation. This is especially crucial for detecting and integrating the relatively broad lines or low-intensity signals in NMR spectra in solutions that can overlap and arise from a variety of sources, including chemical exchange or scalar relaxation. Because multiple-pulse NMR investigations are frequently phase sensitive, specific phase cycling must be used when gathering NMR data. The phase programs used in double quantum ^{13}C–^{13}C NMR investigations need 256 cycles. In addition to eliminating or at least reducing artifact signals that emerge in multiple-pulse studies, such phase cycling approaches simplify phase corrections in the final NMR spectra. By choosing the appropriate coherence paths, particular phase cycling can even suppress unwanted signals; for instance, two spectra obtained with a 180° phase shift might be added to cancel specific lines.

5.11. VARIABLE STATES

Determining the isomeric and conformational states of organic molecules is a crucial part of structural analysis, and this is where NMR comes in. In variable-temperature NMR investigations, these states are typically identified by measuring chemical shifts or spin-spin coupling constants. NOE and residual dipolar coupling measurements can be made for systems undergoing fast interconversions or for frozen conformations. The conformational analysis is the study of molecular conformations and the energy required for their changes. It is a subfield of stereochemistry. This is significant because every conformation has a distinct property, and the physical and chemical characteristics are impacted by their combined attributes. The simple example is the ethane, which rotates around a C–C bond to produce the chemical states. The energy diagram shows that although there are four potential conformations, only one is energetically preferred.

The proton–proton space interactions control this energy preference. The dihedral angles H–C–C–H define the conformations and space interactions geometrically. The interactions and energy are at their highest at an angle near zero. The intramolecular interactions and the accompanying dihedral angles govern the relative stability of isomeric forms. The cause of the magnetic non-equivalency seen in NMR spectra is largely dependent on the sort of symmetry and whether or not molecules have symmetry. The amount of resolved signals and their assignments in NMR spectra, and consequently the molecular structure ascertained by solution NMR, are intimately related to symmetry. The automatic assignment of signals in the $_{1}$H and $_{13}$C NMR spectra of organic molecules by different computer systems is based on the idea of molecular symmetry(Bifulco et al., 2007). The chirality of molecules, also known as optical asymmetry, is one of the various components of molecular symmetry that is essential to molecular identification in many biological and biochemical processes. The chirality in NMR spectra is important. High-resolution ^{1}H NMR spectra should show differences between heterotopic and diastereotopic protons by definition. Actually, their chemical shifts and spin–spin coupling constants ought to differ as well because of their different surroundings. The effect varies with temperature, solvents, chemicals, and nuclei.

The homotopic protons or other homotopic nuclei exhibit the same spin–spin coupling constants and chemical shifts in their NMR spectra. Because of their symmetry, the CH proton and CH_3 groups provide a single signal. The ortho protons of the aromatic ring also produce a single signal because of theirquick rotation around the C–C bond. The phenomena of optical isomerism arises when a carbon turns into an asymmetric center when four of its substituents are different. Chirality is the inability of two optical (mirror) isomers to change into one another. In the presence of polarized (anisotropic) light, these isomers can be distinguished. In the structures, the protons are enantiotropic to the asymmetric carbon. Since the magnetic field is isotropic, unlike polarized light, neither of the isomers will be able to be distinguished in ^{1}H NMR spectra that provide the identical chemical shifts and spin–spin coupling constants.

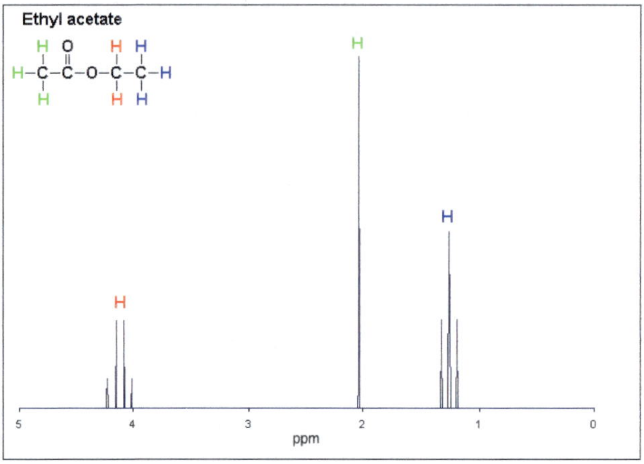

Figure 5.11. ^{1}H NMR spectra.

Source: https://upload.wikimedia.org/wikipedia/commons/1/19/1H_NMR_Ethyl_Acetate_Coupling_shown_-_2.png, Creative Commons License.

5.12. RESONANCE

By detecting two distinct resonances for two chemically identical groups that are adjacent to the asymmetric carbon, one can determine whether a molecule has an asymmetric center in its ^{1}H NMR spectra(Figure 5.11). Even at very fast and free rotation around the C–C bond, the protons that are adjacent to the asymmetric carbon have different surroundings. Two ^{1}H resonances for the CH_3 groups may be visible in a ^{1}H NMR spectrum, with a chemical shift difference that varies according to the compounds, temperature, and solvents. Chirality is associated with a higher number of NMR signals, a feature that frequently led to different interpretations of NMR spectra previously.

Due to the existence of a second asymmetric carbon, which causes diastereomerism, the CHCl protons connected to the asymmetric carbon will be discernible in a ^{1}H NMR spectrum of the molecule. Since the magnetic field is isotropic, chiral solvents, such as optically active amines or alcohols, can be used to provide the diastereotopic conditions necessary for NMR spectrum registration(Wenzel and Chisholm, 2011). The chemical shift differences of enantiotropic protons in these solvents, however, are typically rather minor and vary between 1 and 10 Hz depending on the concentrations, optical purity, temperature, and magnetic field strength. However, it is still possible to identify a compound's optical purity or even its absolute configuration.

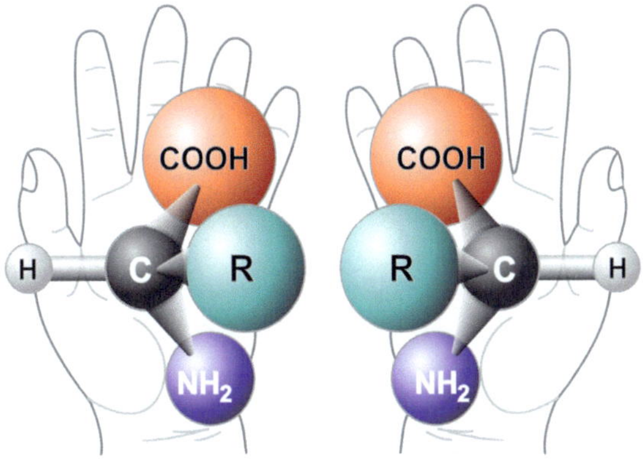

Figure 5.12. Molecular chirality.

Source: https://upload.wikimedia.org/wikipedia/commons/e/e8/Chirality_with_hands.svg, Creative Commons License.

Molecular chirality (Figure 5.12)in solutions can also be investigated by adding paramagnetic optically active chemical shift reagents. These reagents can interact intermolecularly with the coordination centers in molecules under study (oxygen and nitrogen) because they include lanthanide cations Eu^{3+}, Yb^{3+}, or Pr^{3+}. The equilibrium between free and coordinated molecules occurs quickly on the NMR time scale when relatively amounts of the reagents are added, and the resonances that are seen or observed are averaged. The nuclei that are closed centers such as Eu^{3+}, Yb^{3+}, or Pr^{3+} will undergo more paramagnetic oscillations. The molar reagent-to-compound ratios and equilibrium constants, in turn, determine the induced chemical changes. The equilibrium constants are often very large, and a 1:1 ratio is adequate to see obvious impacts. Even complex spin systems ^{1}H NMR spectra can be significantly simplified under these conditions. The fact that Pr^{3+} ions offer negligible Fermi-contact shifts makes them ideal for studies in which resonance lines undergo negligible broadenings. It is obvious that complexation with optically asymmetric compounds in the presence of a chiral shift reagent like $Pr(HFC)_3$ will produce ^{1}H NMR spectra in which the optic isomers will be resolved. Higher chemical shifts, up to 0.5 to 0.7 ppm, will result from the complexation of the shift reagent with (R) isomers of the molecules being studied. It is possible to determine the ratios of the (R) and (S) isomers. The statistically regulated anisochronism of diastereomeric association is a phenomenon seen in enantiomeric mixes. NMR signals that can be utilized for enantiomeric analysis emerge in achiral media that belong to optical isomers as a result of their mutual associative interactions. On the NMR time scale, the rates of these association processes are quite rapid.

CHAPTER 6

MUON SPECTROSCOPY

6.1. INTRODUCTION

A sample can be implanted with the muon, a radioactive particle having a magnetic moment. Due to its radioactivity, the muon has a short half-life. It decays into other particles after an average lifetime of 2.2 μ seconds.

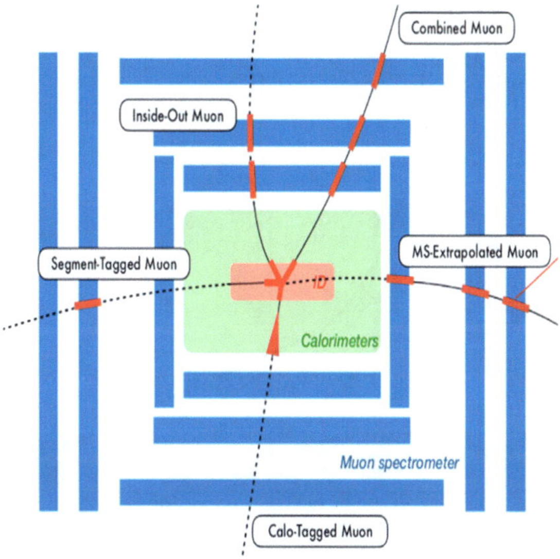

Figure 6.1. Muon spectrometer.

Source: Gazis, E. N. (2024). ATLAS muon spectrometer upgrade for the HL-LHC Era's challenges. *Symmetry, 16*(8), Article 8. https://doi.org/10.3390/sym16081035, Creative Commons License.

We can get the data that the muon probes (Figure 6.1) because of the unique way the particle decays. One subatomic particle that can have either a positive or negative charge is the muon. In μSR measurements, positively charged muons are typically used. The muon has minimal magnetic moment and is a $-\frac{1}{2}$ spin particle. This is not a scattering method, like X-ray or neutron scattering, where an incoming particle is guided in a certain direction toward a sample, followed by a scattering (Parley, 1975). The muon travels in the direction of the sample under study and completely rests or stays inside of it. All of the materials in these tests begin with their spins aligned along a known direction because the positive muons we implant into them are always spin polarized. At a frequency defined by the muon gyromagnetic ratio $\gamma_\mu = 2\pi \times$ 135.5 MHz T^{-1}, the muon spin will undergo Larmor precession in the direction of the local field if a magnetic field is present at the location where a muon stops or halts in a material. A positron e+ and two neutrinos are released when a muon decays (Figure 6.2). The fact that the positron is not emitted isotropically and has a chance of appearing in either direction is a characteristic of this decay Sμ. Rather, at the time of disintegration, it is preferentially released along the muon spin direction. We can determine which way the muon spin was pointing at the time of disintegration by detecting the direction along which the positron is released. Each muon has a different half-life.

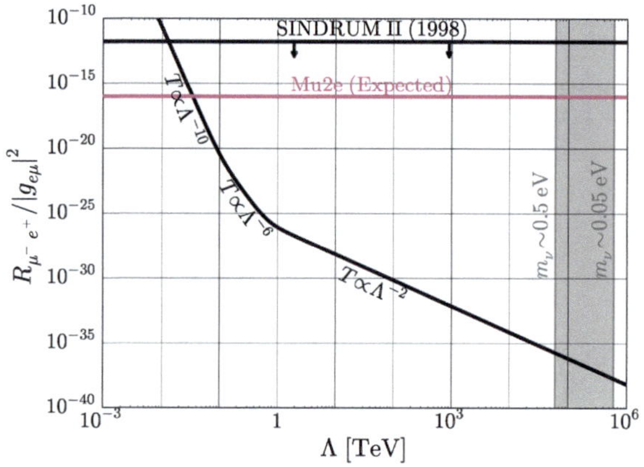

Figure 6.2. Muon decay.

Source: Lee, M., &MacKenzie, M. (2022). Muon to positron conversion. Universe, 8(4), Article 4. https://doi.org/10.3390/universe8040227, Creative Commons License.

Each muon decays to produce a positron, which adds to a data point at a certain time t. A histogram can be created if several positron detections are made across millions of muon decays. The average spin polarization of the vast group of muons as a function of time is then revealed by this histogram. The collection of average spin polarization values as a function of post-implantation time is the data set that we acquire from the experiment or test. The magnetic field at the muon halting or stopping site can be inferred from the polarization P(t) as a function of time t. We can also learn more about the nature of the fluctuations if the magnetic field at the muon site changes over time. The experiments are conducted in establishments that use particle accelerators to create beams. Since the muons are always stopped in the material being studied, μSR observations are not scattering experiments. For the muon to be halted or stopped in a few hundred milligrams of solid matter, the beam energy must be low enough. When the sample is positioned in the beam, the muons stop in the material without causing any damage to the substance or change in their spin orientation. The experiment starts here, with a muon in a known spin state at rest in the sample. One muon at a time is added to the sample.

6.2. MUONS

Muons are an ultra-dilute probe of matter, there are few of them in the sample that there is virtually minimal chance to interact with one another. In order to adjust the sample temperature, it is often attached on a sample stick and put in a cryostat. The muon beam can enter the cryostat through a window. Following muon decay, highly energetic positrons that move at a speed of light readily leave the sample and cryostat. Particle detectors positioned all around the cryostat pick them up. A spectrometer is the device that makes up this array of detectors. The data collecting system, which must receive signals at a very fast rate, and the analysis processors are coupled to the output from these detectors. The lengths of cables and other pathways for electrical communications must be carefully regulated since the timing of the signals must be measured to within a nanosecond (or picoseconds). It is important to keep in mind that light travels at a speed of about 0.3 m per nanosecond. The direction of the initial muon spin polarization S μ (0) is parallel to the application of the magnetic field B. This is referred to as longitudinal field μ SR, or simply LF- μ SR. It is also known as the longitudinal geometry. The direction of the initial muon spin polarization S μ (0) is perpendicular to the application of the magnetic field B. This is called the transverse geometry, and it is commonly abbreviated as transverse-field μ SR, or TF- μ SR(Jayasooriya et al., 2016). However, most spectrometers allow the operator to make the magnetic fieldzero, nulling out the Earth's magnetic field, therefore sometimes studies are conducted with no applied magnetic field. Zero-field SR, or ZF-SR for short, is the name given to this configuration.

The muon provides information about its magnetic interactions at the specific location where it stops in the substance. In a solid, the muon frequently stops in a position that is peculiar to crystallography. There will probably be a variation in the behavior examined at each muon location as we measure the outcome of millions of muon implantations. All of the muons that have been implanted in the sample and have interacted with it throughout that period t are averaged when we measure the polarization at each time t. The 'relaxation of the polarization' is the phrase used to describe the polarization decay. The precise form that this relaxation takes contains information that can be accessed to gain a better understanding of electrical characteristics, magnetism, superconductivity, and material chemistry. Carl David Anderson and Seth Neddermeyer discovery of the muon in California marks the beginning of the muon evolution. These scientists used the tracks left in a cloud chamber to analyze powerful cosmic rays. The density of condensed water droplets in the tracks can be used to calculate the particle velocity, and the magnetic field curvature in such tracks can be used to calculate the ratio of the momentum of the particles generating them to their charge. The mass-to-charge ratio can be inferred from these two characteristics. The positron was discovered by Anderson using the same method. The mass of the recently discovered muon particle is approximately 200 times that of an electron and 9 times that of a proton. The muon, denoted by the symbol μ, can be positively or negatively charged. The muon was first classified as a meson because of its mass, which is in between that of the light electron, which is classified as a lepton, and the heavy proton, which is classified as a baryon. In 1935, Hideki Yukawa predicted that the particle causing the strong nuclear interaction would be the meson. Of all the unstable elementary particles, the muon has the longest lifetime. This characteristic provided an early illustration of the application of the muon in answering a scientific question by enabling the first experimental confirmation of the time dilation effect predicted by Einstein's special theory of relativity. Bruno Rossi and David Hall conducted this investigation.

6.3. OBSERVATIONS

To study the count rates for energetic muons traveling vertically downstream at various elevations, they traveled to Colorado in 1941. They discovered that, in accordance with Einstein's prediction, the lifetime of the muons rose with the muon momentum when they compared the rates in Echo Lake (3,240 m) and Denver (1,616 m). They came to the conclusion that the muon's lifetime in its inertial reference frame was 2.4 μ s, which is in line with the more precise value that we know now. The Standard Model classification of fundamental particles serves as the foundation for particle physics theory. The muon is an elementary particle in this classification rather than a meson. As two leptons in the six-membered lepton family, the muon and neutrino are now in existence. The electromagnetic interaction affects the muon. The Standard Model classification of elementary particles includes the muon as one of six leptons(Grupen, 2005). Following the proposal of T. D. Lee and C. N. Yang, C. S. Wu researchers found a key property of the weak force in 1957.

This feature is that weak decay processes do not preserve parity, or spatial inversion symmetry. The negative helicity of neutrinos and the asymmetric positron emission that supports μ SR are caused by this parity non-conservation characteristic. The fact that almost all muons follow this decay path is another significant aspect. The muons are an extremely effective source of the elusive neutrinos. To determine mass, these neutrinos can be steered toward specialized detectors located at distant locations. Therefore, high intensity muon beams can be used as neutrino factories to aid in the investigation of important cosmological issues like the genesis of dark matter.

Figure 6.3. Large hadron collider.

Source: https://upload.wikimedia.org/wikipedia/commons/4/46/Views_of_the_LHC_tunnel_sector_3-4%2C_tirage_2.jpg, Creative Commons License.

A primary focus of high-energy particle physics research is currently the Large Hadron Collider (LHC)(Figure 6.3)at CERN, which famously produced the first experimental confirmation of the long-predicted Higgs boson in 2012. Subsequent research is still being conducted to determine additional characteristics of the Higgs particle and its interactions. Tracking high energy muons is a crucial component of particle physics research since they are significant decay products of the Higgs particle. Muon collider concepts are being intensively pursued in particle physics as it looks to expand the energy frontier beyond the LHC. When it comes to developing higher energy accelerators for future research, muons have a significant advantage over hadrons because they are primary particles and lose less radiative energy than electrons. The deviation of particle properties from the exact predictions of the Standard Model is another scientific area that muons can address. Significant deviations of the muon 'g' factor have been found by the muon g–2 experiments, and the experiment precision is constantly being increased.

Figure 6.4. Muon spectroscopy.

Source: Biswas, S., Gerchow, L., Luetkens, H., Prokscha, T., Antognini, A., Berger, N., Cocolios, T. E., Dressler, R., Indelicato, P., Jungmann, K., Kirch, K., Knecht, A., Papa, A., Pohl, R., Pospelov, M., Rapisarda, E., Reiter, P., Ritjoho, N., Roccia, S.,&Amato, A. (2022). Characterization of a continuous muon source for the non-destructive and depth-selective elemental composition analysis by muon induced x- and gamma-rays. *Applied Sciences, 12*(5), Article 5. https://doi.org/10.3390/app12052541, Creative Commons License.

6.4. MUON SPECTROSCOPY

The two main characteristics are the asymmetric angular distribution of the released positrons with respect to the muon spin direction at the moment of decay and the 100% spin polarization of muons generated by pions decaying at rest. Since they provide access to the muon spins initial and final directions and, consequently, its time evolution, both are crucial for the SR experiment. Muon spectroscopy (Figure 6.4) is a method for studying low energy processes in muon-containing materials. This method is also known as μ SR, where μ S denotes muon spin and R can be interpreted as representing one of the three primary forms of muon spectroscopy measurements: rotation, relaxation, or resonance. One can simply assume that the R in μ SR stands for research in general. In order to conduct μ SR studies, the muons must be slowed down to thermal energies during the implanting process in the sample of interest. Athick sample can be positioned inside the muon beam to do this, or the beam can be slowed down or degraded by placing material in front of the sample. Two essential aspects of the implantation procedure are quick thermalization and muon polarization preservation. Since the thermal energies associated with the sample are usually on the meV scale, the muon must sacrifice a significant amount of energy. Positive muons are frequently utilized here, and the typical muons used in muon spectroscopy start off with an energy of 4 MeV. Atomic ionization and excitation activities initiate the muon energy loss, which then moves on to charge exchange, capture, and loss processes. Monte Carlo techniques, which are helpful when creating tests, can be used to model the stopping process. The phrases diamagnetic and paramagnetic, which describe the magnetic properties of the electrons to which the muon is associated are frequently used to classify muon states. The muon can be described as existing simply as $\mu +$ or Mu+, or it can be coupled to two electrons that are in a singlet state or it can be uncoupled to any unpaired electron density, which results in a diamagnetic muon state(Fleming et al., 2012). When a muon is linked to a single unpaired electron, as in neutral muonium, a paramagnetic muon state is created. The sort of sample under investigation will determine the nature of the final muon states that are obtained in the sample. The final muon state in metals typically stays in diamagnetic, positive muon state. One possible state in insulators is the diamagnetic positive muon state, but there is also the paramagnetic state, which retains an unpaired electron. Muonium, a muon counterpart of the neutral hydrogen atom, is the most basic type of paramagnetic state. The response of these various states to a transverse field (TF), or a magnetic field applied perpendicular to the initial muon polarization, is what distinguishes them. It should be mentioned that states exhibiting this diamagnetic response could potentially be the consequence of a fast-changing paramagnetic local environment that exhibits a significant shift in relation to the bare muon. Much of the muon spectroscopy work done at the several muon facilities worldwide is based on a condensed matter physics. The thermalized, implanted muons serve as local indicators of both static and dynamic magnetism in the studied sample. Complementary site computations can then be developed to support the analysis of the data in relation to the specific sites. Superconducting vortex lattices with an extended length scale in relation to any individual locations inside the lattice would be an appropriate example. This category may also include magnetic systems with a high degree of disorder or those with a large cell and several stopping locations. The muon can be used to investigate the fluctuating magnetic fields linked to different dynamical processes in the sample in addition to studying static magnetism. It may be observed that the μ SR approach offers a link between generally slow techniques like magnetic remanence and a.c. susceptibility and characteristically quick techniques like neutron scattering. There are various benefits when comparing μ SR to bulk magnetic investigations. Phase-segregated samples can be readily identified in muon data, while μ SR, being a volume-averaged probe, is comparatively resistant to low concentration impurity phases that can dominate bulk susceptibility readings at low temperature. Additionally, μ SR enables measurements in zero applied field and is especially suited for researching antiferromagnetism. The remarkable sensitivity of μ SR to extremely small magnetic fields and moments, it can usually detect fields below 10 μT and moments below 0.01 μB, is another of its advantages. Because the entering muons can pass through the cryostat windows and the exiting positrons can pass through the cryostat walls and windows, the method works very well at extremely low temperatures(Fleming et al., 2012).

6.5. μ SR

From a different angle, μ SR is seen as a part of the larger family of magnetic resonance methods. Unlike the simple thermal polarizations found for NMR nuclei, it has 100% spin polarization. The muon has a higher gyromagnetic ratio than any other nucleus. Since muons may measure in any substance, there is no need for particular nuclear isotopes. In the current generation of high field muon spectrometers, the ability to employ an extended field range for measurements, from zero field (ZF) up to 10 T. They exhibit adaptability to sample conditions with low temperatures such as helium dilution refrigerators. The short muon lifetime limits the ability of RF pulse sequences to manipulate spin. To properly understand the data, the muon halting or stopping sites must be identified. The presence of the muon will cause some disruption of the system being studied.

The understanding that the chemical characteristics of Mu are similar to those of the H atom, the most basic type of free radical and therefore highly chemically reactive, forms the basis of the chemistry standpoint. A highly reactive light isotope of H with a radioactive label is provided by muonium. By using muon spectroscopy to study Mu, it is possible to track the transient starting stages of processes and identify the time dependence of chemical reactions involving a hydrogenic species. The hydrogen isotope sequence of H, D, and T is significantly extended by the presence of Mu. Muonium chemistry also has the advantage of having a large quantum zero-point energy due to its low mass, which allows it to probe the surface of the interatomic potential energy surface (PES) and offer a comprehensive test of chemical reaction theory. The muon can be raised even farther up the PES i.e., at least three times higher than the ground state by vibrational excitation, such as with a laser. The implanted muon problem is a variant of the hydrogen interstitial defect problem (computational science). The structure of the interstitial muon defect, including the atomic relaxation surrounding the muon site, can be determined using computational techniques like density functional theory (DFT)(Huddart et al., 2022). Since typical computational approaches like DFT only supply the classical site, the muon strong quantum character presents a hurdle for computational investigations. To accurately represent the muon's quantum delocalization, more computing steps are needed. A spin-polarized muon is stopped in a material at the start of the μ SR experiment. Mu^+ is the muon is a positively charged particle that can exist independently. This is comparable to the hydrogen cation H^+. Although $^2H^+$ is commonly referred to as hydron, it is actually just a proton p^+. Physicists are more inclined to write the species as μ^+. Because they are familiar with the analogy with hydrogen, chemists typically write Mu^+.Mu^0 can absorb an electron to create the electrically neutral bound state μ^+e^-. The hydrogen atom H is comparable to this. It is in the neutral, paramagnetic condition, we can write Mu as Mu^0. Keep in mind that although H_2, or molecular hydrogen, is the chemically stable form of hydrogen, the muon flux is never big enough to create Mu_2. Similar to the hydride ion H^-, the muon can acquire two electrons to create the negatively charged bound state Mu^-. This state is an electron singlet, or diamagnetic. These several charge states, which are all influenced by the electronic structure surrounding the muon, can be viewed as separate chemical species. However, Mu^+, Mu^0, and Mu^- are perfect states. The final state may be weakly coupled to nearby atoms or there may be screening effects in a metallic sample even after full muon thermalization. This implies that the simple images connected to these idealizations may not accurately reflect the actual spin and charge concentrations. However, $Mu\pm$ and Mu^0 are helpful even in unsaturated molecules and magnetic materials, where the spin density becomes increasingly delocalized. It is crucial to remember that the muon is both a magnetic and a chemical species. The muons are injected at energies of 4 MeV. Ionization of the sample atoms is the main energy loss mechanism that causes the muons to decelerate down. Once the muon kinetic energy falls below the ionization energy, which is usually of order 10 eV, the charge cycling will cease, and the remaining energy will be dissipated as phonons. Because the Coulomb interaction does not couple to its magnetic moment and the period spent as a neutral atom is far too short for the hyperfine interaction to have any effect, the muon spin remains basically unaffected during the charge cycling and slowing down. With a specific (material-dependent) chance or probability of possessing an electron, the muon then settles slightly downstream of the ion track, creating neutral Mu^0. The muonium (Mu^0) electron affinity is 0.75 eV. This is the energy required to extract one of the electrons from Mu^- and convert it back into Mu^0. It should be noted that Mu^0 is more stable because it requires 13.6 eV to ionize to Mu^+.

6.6. TRANSITIONS

After stopping, the muon may have quick charge-state transitions or reactions on a period of a few nanoseconds. We can observe its spin relaxation with any experiment data acquisition. Mu^+ will look for an area with a high electron density, like a bond-centered site, a lone pair on an atom (like O), or near an electronegative ion (like F). The hydronium ion H_2MuO^+(H_3O^+) can be formed by attaching to a water molecule, or it can bind to an oxide ion to generate a single negative hydroxide ion. Mu^+ can also quickly form a spin singlet bond in covalent materials, indicating a spin-unpolarized screening charge. Usually, this is true for non-magnetic metals. The neutral atom may react with a nearby molecule in molecular liquids and solids, maybe targeting a double bond or lone pair site. As an alternative, it might locate an interstitial space, where it can continue to exist as a free atom (intrinsic semiconductors). Charge state changes in some instances. This could happen if the muon manages to grab an electron from the substance, possibly one of those released along the stopping track. If suitable, an electron can also be returned to the material (conduction band). The muon will dephase in transverse field if it remains in the neutral (paramagnetic) state for a duration equivalent to the inverse hyperfine constant before changing into diamagnetic Mu^+ or Mu^-. Since the crucial timescale is the period of the triplet precession, the phase shift and amplitude will be field-dependent for low field when the Mu^0 proportion would give triplet precession. If the muon electron is connected to adjacent nuclear spins, the amplitude in zero field will decrease. If Mu^+ is slower to convert to Mu^0, similar dephasing is seen in the transverse field (Fleming et al., 2015). Energy barriers may slow down reactions and charge exchanges. It will take time for electrons or holes from the muon track to decay to the muon. These modifications typically lead to muon polarization relaxation, which is typically exponential in shape. There is no instantaneous change in polarization in longitudinal field when Mu^0 transforms into Mu^+ or Mu^-. Techniques like muon spin resonance with a delayed RF pulse can still be used to detect the change. With the same time constant as in low longitudinal field or transverse field, the conversion of Mu^+ to Mu^0 in high longitudinal field slightly alters polarization. Higher temperatures can cause charge cycling, for instance in semiconductors where there are enough electrons and holes for interaction. This relaxation will still be seen as exponential. A field scan will provide relaxation to zero polarization at any field, since it will take many charge cycles to fully depolarize the muons at high longitudinal field. Given that spin exchange may occur whenever charge exchange is effective, analysis of this type of data can provide the charge cycling time and relative instantaneous occupation of the various states. Metals are unique among all materials because none of them contain muonium (Mu^0). Rapid spin exchange guarantees that any hyperfine interaction is averaged and that the muon seems to be in a diamagnetic state, such as Mu^+, because the electron orbitals of the muon overlap with those of the conduction electrons. Although there may be local distortion and charge redistribution near the muon site, the conductivity should guarantee that the muon point charge is effectively screened on a longer scale. In order to evaluate the results of a sample, we must know the muon charge and spin state or states, particularly if the muon is meant to monitor characteristics like magnetism or ionic mobility. Observing the magnitude of the Larmor precession for the bare muon at low transverse fields (2 mT) is method of verification(Jegerlehner and Nyffeler, 2009). Paramagnetic Mu^0 must be represented by any missing amplitude, either as the free atom, bound in a radical, or as a missing fraction that was first produced as one of these but changed to a different state. A triplet muonium precession signal can be seen from interstitial Mu^0 in favorable circumstances, which are usually samples with few or no nuclear moments. Usually, a pulsed muon source with a frequency of about 2.8 MHz uses 0.2 mT. Low transverse fields can show phase and amplitude variations. The initial state mean lifetime can be computed from this. The repolarization of the radicals or muonium is provided by a longitudinal field scan. In order to determine the proportion in that state plus the approximate hyperfine constant, scans are usually conducted from zero to about 0.2–0.5 T. Any missing fraction is also included in repolarization. Anisotropy or dynamics will have an impact on the curve shape. Since neither Mu^+ nor Mu^- possesses an electrical moment, it is more difficult to distinguish between them from the muon data alone.

6.7. COUPLING

The Mu^+ may exhibit a nuclear dipole coupling in zero field since it will be bonded closer to other atoms. In oxides where the proportion of magnetic ^{17}O is relatively small and a Mu^+ is likely to have oxygen at its vicinity, this is not a reliable indication. By altering a semiconductor doping and tracking a shift in the diamagnetic fraction, one might infer the state that is present. A more n-type sample will typically favor Mu^- over Mu^+.There is a magnetic moment in the muon. A classical magnetic moment can be compared to a current loop. The magnetic moment μ is defined by the current *I* multiplied by the area of the loop A when a charged particle travels in a closed orbit, creating a current loop. Note that a charged particle also has mass. Not only does charge travel in a circular motion, indicating a magnetic moment, but mass also moves in a circular motion, indicating angular momentum. This indicates that the orbital magnetic moment μ and the orbital angular momentum *L* are related and ought to be proportionate; the gyromagnetic ratio is the proportionality constant. Therefore, $\mu = \gamma$ L. It should be noted that the negatively charged electron e^- has a negative magnetic moment and gyromagnetic ratio because we can work with the positive muon μ^+, which has a positive magnetic moment and gyromagnetic ratio. The fluorine ion is rather special because it is highly electronegative and therefore attracts the positively charged muon, the ionic radius of F^- is very small, allowing the muon reside close, the nucleus of fluorine has spin one-half, allowing any relaxation to retain a distinctive time dependence, and the nucleus has a large nuclear moment, linked to an isotope that occurs with 100% abundance. It was acknowledged that a muon stopping between two fluorine ions to create what is known as a F–μ–F state may result. This is significant because the polarization function that characterizes this time evolution is highly dependent on the type of material our muons are implanted into, and the polarization function is precisely related to the asymmetry A(t) that we measure in a μSR experiment. The fundamental cause of the time evolution is that a single spin would precess in a magnetic field that is perpendicular, causing the muon spin polarization to oscillate. The oscillating signal may be eliminated by averaging across a large number of muon precessions, which would cause the polarization to relax. The muon time in the system often results in changes to the magnetic field at the muon location. When dynamics are present, the polarization function is derived under the assumption that the dynamic process is a stationary Gaussian-Markovian process. This suggests that the dynamic characteristics are unaffected by the time evolution. The field distribution is available at any one time is Gaussian. Pα(t) is frequently derived using two models. In the first, B(t) undergoes a sequence of abrupt changes, but in the second, it is assumed that the field changes gradually throughout time.

Both approaches can be used equally for the derivation of Px(t) and Pz(t) and produce relatively similar results. Because B(t) abruptly shifts from one value to another, the first model for calculating Pα(t) is frequently referred to as the strong-collision model. The muon ensemble in this model changes over time in a static field. The local field experiences a discontinuous fluctuation event that modifies its value after an average time τ. By rearranging the magnetic moments, it selects a new value based on the range of fields that are accessible. After that, the muon ensemble continues to exist in the new static configuration until experiencing yet another abrupt transition. The muon precession process is effectively reset on a regular basis. Dynamicization is the act of accounting for these processes, and it can be done quantitatively. Muon spectroscopy is a valuable tool for examining condensed matter magnetism(Blundell, 1999). The muon spin is a sensitive probe of small magnetic fields due to the magnitude of the muon gyromagnetic ratio, $\gamma\mu = 2\pi \times 135.5$ MHz T^{-1}. In magnetically ordered materials, the muon spin might react to the presence of normal magnetic fields because of the muon lifetime of 2.2 μs. It is crucial to note that this behavior is just controlled by a power law; no specific length or time scale is needed. The power-law behavior of thermodynamic quantities close to a phase transition is described by the Landau model. When conditions are favorable, μSR can be very effective at measuring critical exponents. For example, it is frequently preferable to lock onto a precession signal and measure its frequency as $M \rightarrow 0$ as $T \rightarrow Tc$ rather than trying to measure the amplitude of a weak magnetic diffraction peak as it fades into the background noise.

Two interpenetrating ferromagnetic sublattices, one with all spins pointing up and one with all spins pointing down, can be conceptualized as a result. When there is no applied field, the most basic μSR experiment on a magnetically ordered material occurs. The local microscopic spontaneous magnetization M is then proportional to the local magnetic field at the muon site, which is mostly governed by the dipolar field of the ionic magnetic moments. The contribution of the muon spin component parallel to the field that does not precess is captured by the $\cos^2\theta$ term. The component of the muon spin that is perpendicular to the local field is captured by the $\sin^2\theta$ term. This component must precess at a frequency $\omega = \gamma_\mu B$, which is dictated by the strength of the local magnetic field. By this precession, we can determine the strength of the local field at the muon site, and the amplitudes of the contributing components encode information about the field's direction. The polarization of many muons is typically measured in an experiment as a function of time, as an average across the suitable distributions of the field direction $f(\theta, \varphi)$ and the field magnitude p(B), which characterize the muon ensemble magnetic environment. We estimate the muon ensemble average polarization Pz(t) and its time evolution. The situation where a magnetic field points along every possible direction with an equal probability with respect to the initial muon spin direction is an illustration. This could be used to characterize a polycrystalline, completely magnetically ordered material where the muons always perceive the same strength of the local magnetic field, but each of the many crystallites in which the muons implant has a different magnetization direction. $Cu(pyz)_2(ClO_4)_2$ serves as an example of how the presence of several muon sites can yield valuable information(Schlueter et al., 2012). With the magnetic moments on the Cu^{2+} ions, this molecule is a molecular antiferromagnet that exhibits a complex time-dependent muon spectrum at low temperatures. Three distinct frequency components can be identified by Fourier transformation of these data. The best fit, represented by the solid line through the data, is obtained by fitting the data in the time domain using three frequency components. Despite being sized differently, these three curves all exhibit the same temperature dependence. It can be inferred that each frequency has a separate constant of proportionality that reflects the location of the three distinct muon sites with respect to the copper moments, but is proportional to the same order parameter. The frequencies at various muon sites may exhibit distinct temperature dependences. Since we see a precession frequency proportionate to the order parameter in both ferromagnets and antiferromagnets, the muon data for these two types of particles appear to be rather similar. Therefore, it is generally not possible to clearly and consistently differentiate between ferromagnets and antiferromagnets using muons.

6.8. MAGNETIC SUSCEPTIBILITY

Having access to the temperature-dependent static magnetic susceptibility $\chi(T)$, (Figure 6.5)which typically provides a clear indicator of whether a material hosts ferromagnetic or antiferromagnetic interactions, is always beneficial as μSR is not a technology in isolation.

Susceptibility is a bulk thermodynamic method that truly quantifies a sample total magnetic moment, or $\int MdV$, where V is the volume, when a field H is applied to the entire sample(Parks, 2018). It can only provide a clear answer if M is constant across the sample, making it unable of differentiating between a sample with a small quantity of a strongly magnetic impurity and one with a homogeneous, weakly magnetic response. The muon asymmetry will locally examine whether a material contains distinct phases. Their relative amplitudes will show the proportions of the magnetic phases present, giving them the appearance of separate spectrum components. A material is said to display magnetic dynamics if its minute magnetization varies with time. These dynamics have an impact on the muon polarization in magnetic materials. We frequently address two closely related ideas or concepts: dissipation, which is the loss of energy in a system as a result of the generation or extinction of excitations, and dynamic fluctuations, which are the time-dependent departures of the microscopic magnetization from its average value. Both of these cause the time-dependent muon polarization to relax. The correlation function and the dynamic susceptibility are two interesting functions that can be obtained experimentally. The dynamic susceptibility is proportional to the dissipation, correlation functions characterize the fluctuations. Since the fluctuation-dissipation theorem closely connects the two ideas,these two values actually measure almost the same thing.

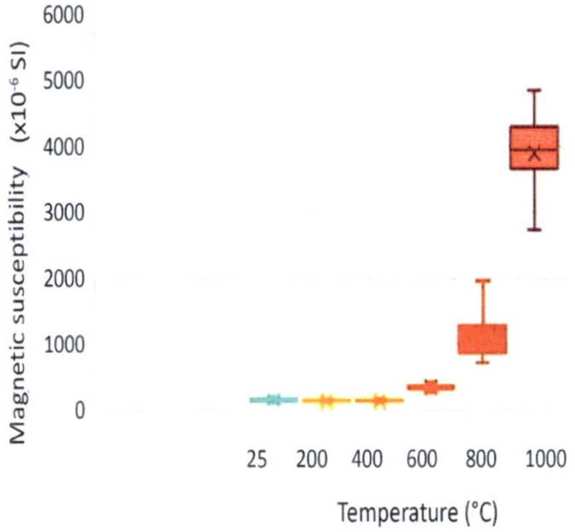

Figure 6.5. Magnetic susceptibility.

Source: Sanchez-Roda, A., Oliva-Urcia, B., & Gomez-Heras, M. (2022). The use of magnetic susceptibility as a technique to measure the impact of wildfires on archaeological heritage. *Applied Sciences, 12*(19), Article 19. https://doi.org/10.3390/app121910033, Creative Commons License.

6.9. SPECTRUM

The excitation spectrum is altered when symmetry breaking occurs. This is due to the emergence or development of new, slowly changing quantities called broken-symmetry variables. A new excitation in the magnetization, the spin wave, can be supported once symmetry in a ferromagnet on magnetic ordering is disturbed. The eigenstates of the magnet in the limit $T \to 0$ are known as spin waves, or magnons. Each one represents a propagating mode, which is a wave-like excitation in the magnetization that differs qualitatively from a diffusing mode. For larger T, where multi-magnon processes occur, magnons are less well characterized and ultimately degenerate into the diffusive behavior. According to energy conservation, the degree to which the characteristic frequency ω coincides with the muon precession frequency ω0 = γμB is a critical factor in any connection between an excitation of lifespan 1/ω and the muon spin. The exchange, J/h, sets the top limit of the frequency range of spin waves at the Brillouin zone boundary, which is often substantially more than ω_0. The bottom limit for acoustic modes in isotropic magnets with q → 0.15 is ω = 0. High temperatures, where diffusive excitations predominate and magnons are less strongly defined, are necessary for this second-order process, which calls for a huge population of magnons. A system with comparatively fewer magnetic moments than lattice sites is said to have a diluted spin. These moments are magnetically disordered i.e., the spin direction is randomized and organized randomly on the lattice. We permit time-dependent fluctuations in the disordered moments. This is the case with spin glass materials, which are made up of a small concentration of magnetic ions dissolved in a metallic, non-magnetic material. Examples include Mn dissolved in Cu at a concentration of around 1% and Fe dissolved in Au.

The magnetic exchange between the magnetic constituents is mediated by the RKKY interaction. However, the exchange interaction between the ions is highly randomized due to the oscillatory nature of RKKY exchange in space, and its amplitude and sign fluctuate depending on how far apart the magnetic ions are. In such a system, we can attempt to calculate the muon response. Every exponent may have a relationship to the static behavior near phase transitions. A gas or cloud of electrons interacting through the Coulomb interaction is known as a metal. This interaction gives metals a structure of excitations and characteristics, such as a magnetic response. Finding extra constraints, like those on molecular reorientation via avoiding level crossing spectroscopy, is to lessen correlations between parameters. The ring-structured single molecule magnet Cr_7Ni and its Cu bridging dimer Cr_7Ni-ent serve as an example. They are medium-sized molecules with many more hydrogen atoms and tens of oxygen atoms. At varying separations from the magnetic ions, they produce several muon sites. Their LF relaxation rate makes it possible to identify the muon sites that are farther away from the magnetic ions. There is very limited intermolecular interaction when the molecules form crystals. A magnetic ground state and many excited states are produced via intramolecular exchange, which is subject to multiple relaxation channels for the ensemble magnetization. Sharp BPP-like peaks are visible in both molecules, and the distinct field dependency enables a self-consistent assessment of the model parameters. However, the employment of more intricate models with activation temperature distributions is necessitated by the presence of more relaxation channels. This occurs frequently when the BPP method is used. Condensed matter also contains these mobile excitations more generally, and μSR can be used to directly examine their characteristics, especially in low-dimensional systems. Unlike wave-like excitations like phonons and magnons, which are localized in momentum space and extended in real space, these mobile excitations can be conceived of as localized in real space. A basic anisotropic lattice model offers a common method for comprehending how mobile excitations in a material might be probed using static implanted muons. By adjusting the anisotropy of these rates, a system with decreased effective dimensionality can be addressed. The hopping rates of the mobile excitations along the three main axes of a cubic lattice are assumed to be D1, D2, and D3(Grillet et al., 2011). A muon positioned at a lattice interstitial site will couple to fluctuating fields linked to the excitation mobility. When the excitation approaches the muon site, there may be a dipolar coupling D as well as a contact hyperfine coupling A that interact directly with the mobile excitation. An indirect approach, in which the confined electron interacts with the mobile electron spins through an intermittent spin exchange process, is feasible if a muon is linked to a localized electron in a muonium-like probe state. The external electron spin fluctuations will thereafter be transferred to the muon by the internal hyperfine coupling within the muonium state that is defined by A and D.

The excitation types may represent unpaired electron spins. Similar considerations might be made for other excitation types with corresponding magnetic fields, such as barriers and other topological excitations in magnets or charge carriers and triplet excitons in semiconductors. It is not even strictly required for the mobile excitations in the muonium-like probe state to have a magnetic character because a spinless structural excitation that modifies the muonium hyperfine coupling (A,D) will also result in an effective modulation of the local field at the muon site, which will cause muon spin relaxation. The conformational excitations of a polymer chain may serve as an example. Researchers observed that the relaxation rate seemed to decrease with time, indicating that the time evolution of the relaxation was non-exponential, despite the fact that an exponential was initially utilized to match the relaxation. Initially, this was understood to mean that the interchain motion of solitons became increasingly significant over longer periods of time. But since the correlation function for 1D diffusion is not exponential, the muon spin relaxation is also not anticipated to be exponential.

6.10. RISCH AND KEHR(RK)MODEL

RK specifically examined this problem and noted that the standard NMR theory used for the earlier data analysis is not strictly valid because it assumes the existence of a finite correlation time that is short in relation to all other timescales in the problem, whereas the correlation time for the return to the origin of a particle diffusing in 1D is divergent. For the case of a static muon interacting with an electron that is randomly diffusing along a 1D chain via an intermittent hyperfine connection, RK subsequently created a stochastic diffusion theory. The inadequacies of the NMR method for muon-spin relaxation were shown by the RK model. The perturbation theory is a useful place to start, however it does not always accurately explain the evidence from muon experiments. The main reason for this is that the NMR theory relies on frequency and timescale assumptions that are not always suitable for explaining μSR experimental settings. Although the exact form of the relaxation cannot be precisely described by the exponential relaxation model of perturbation theory, it is important that in practice, a relaxation rate that scales with

$B -\frac{1}{2}$ can be obtained by fitting data described by the RK model to an exponential function. In certain situations, the muon itself moves rather than a stationary muon that follows the dynamics of its surroundings. Phonons aid in the muon's hopping between interstitial sites, which causes it to increase with temperature(Karlsson, 1982). Activated rates are roughly proportional to

$T-\frac{1}{2}\exp(-Ea/kBT)$,

where Ea is an activation energy. This happens because the muon is first self-trapped by its own local distortion of the lattice, and a tunneling transition can only take place if two adjacent energy levels coincide due to the thermal fluctuations of the lattice caused by phonons. After tunneling through the barrier, the muon can become self-trapped in the subsequent location. The hop rate drops when the sample cools because there are less phonons. But when the very low temperature zone is reached, when the phonons seem to be impeding rather than facilitating hopping, the copper example reacts differently. The muon begins to delocalize into a wavelike form at low temperatures due to the possibility of coherent tunneling. The coherence of this delocalized state is now destroyed by inelastic scattering, which is caused by phonons. Quantum diffusion is the term used to describe this coherent process. Note that the kind of quantum diffusion described above won't happen in many materials because the muon creates a chemical connection with the structure (oxides). The muon might, however, be thermally stimulated out of this bonded site and jump to an adjacent one at high temperatures. When the energy barrier between a close pair of sites is not too high, a more localized type of quantum diffusion in addition to long-range quantum diffusion is feasible. The quantum tunneling proton dynamics and hydrogen bonding that occur in many materials are comparable to the resulting quantum tunneling between local sites. In comparison to a proton, the muon's low mass increases the tunneling rate and the bonding effective range. Estimating the muon halting sites may become more difficult as a result of this quantum tunneling across groups of local potential minima.

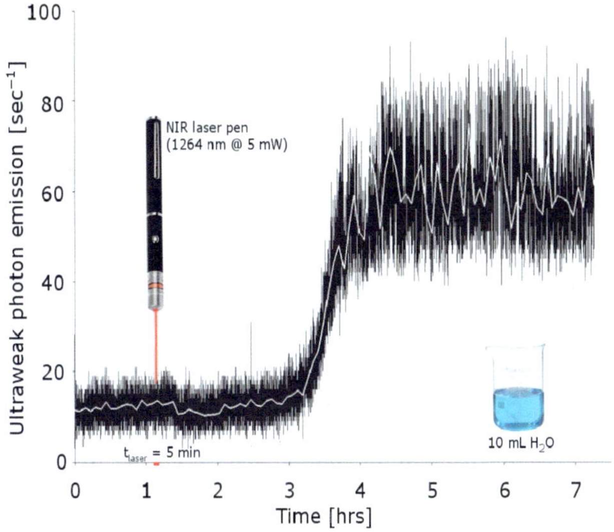

Figure 6.6. Quantum coherence.

Source: Madl, P., &Renati, P. (2023). Quantum electrodynamics coherence and hormesis: foundations of quantum biology. *International Journal of Molecular Sciences, 24*(18), Article 18. https://doi.org/10.3390/ijms241814003, Creative Commons License.

6.11. QUANTUM COHERENCE

The presence of macroscopic quantum coherence (Figure 6.6)in the superconducting state is already implied by zero electrical resistance. Decoherence is almost absent for superconductors because electrical resistance disappears when τ is infinite. According to quantum band theory, solid-state electron energy levels are arranged into quasi-continuous energy bands and divided by energy gaps. When an electric field is applied, partially filled bands that permit electron motion are what define metallic behavior. Since fully occupied bands define the behavior of semiconductors and insulators, electrons must have an energy of the order of the bandgap energy, for example, about 1 eV for silicon, in order to move from the lower (full) valence band to the upper (empty) conduction band and subsequently contribute to conductivity. This can only be accomplished by thermal excitation at high semiconductors reflect extra levels brought into the bandgap by impurities and defects, or by stimulation by light with energy greater than the bandgap (photoconductivity). These levels can be thermally excited from the lower valence band, leaving an empty level in the valence band (hole), or to the upper conduction band, producing n-type conductivity, using a far lower energy than the bandgap. In the latter scenario, p-type conductivity is caused by holes, which can be represented as positively charged quasi-particles. The impurities and defects are crucial to semiconductor physics. The location of the defect level in the bandgap, which distinguishes between shallow and deep impurities, is one of the crucial factors. Shallow levels are particularly useful for n-type or p-type doping because the defect level is extremely near to the bottom of the conduction band (shallow donor) or the top of the valence band (shallow acceptor). A superposition of conduction band or valence band states that result in hydrogen-like levels can sufficiently explain the electronic wave-function of the shallow defect. Usually, the deep levels in the mid-gap are efficient recombination centers, sometimes referred to as traps. Certain impurities, like the Si impurity in GaAs, can exhibit several configurations in the semiconductor; if both donor and acceptor configurations are feasible, the impurity is said to be amphoteric. A material may contain multiple impurities or defects. Electrically active impurities react to form a complex without a defect level in the bandgap, eliminating the electrical levels of the isolated impurities. The impurities stay isolated from one another but have opposite effects in electrical conductivity (donor vs acceptor). Of all the contaminants, hydrogen is very important. Protons can be introduced into semiconductor materials and devices in a variety of production phases. Hydrogen is ubiquitous and can occasionally reach quantities of up to 1%, which can significantly impact the material characteristics. One example is the hydrogen passivation of hanging bonds in amorphous silicon, which eliminates the associated trap levels from the bandgap and improves the efficiency of the corresponding photovoltaic systems. One of the important steps in the creation of blue LEDs was determining the function of hydrogen in the passivation of magnesium acceptors in GaN(Zhang et al., 2020). Describing the isolated impurity and its potential interactions with other impurities and defects at the microscopic level is a crucial first step in comprehending hydrogen in semiconductors.

Muon spin spectroscopy (Figure 6.7) has helped us to understand semiconductor physics. Since hydrogen is a difficult atom to detect experimentally and usually requires high concentrations, researching muonium as a hydrogen analog has shown to be a very successful strategy. Other than μSR, only a few systems, like ZnO and TiO_2, where hydrogen is often present in significant concentrations can employ experimental approaches sensitive to isolated hydrogen itself. It is more challenging to get microscopic data regarding the electronic structure, electronic levels, and hydrogen structures. Since muonium is a pseudo-isotope of hydrogen, the majority of the information currently available regarding isolated hydrogen in semiconductors and dielectric insulators has been obtained from μSR.

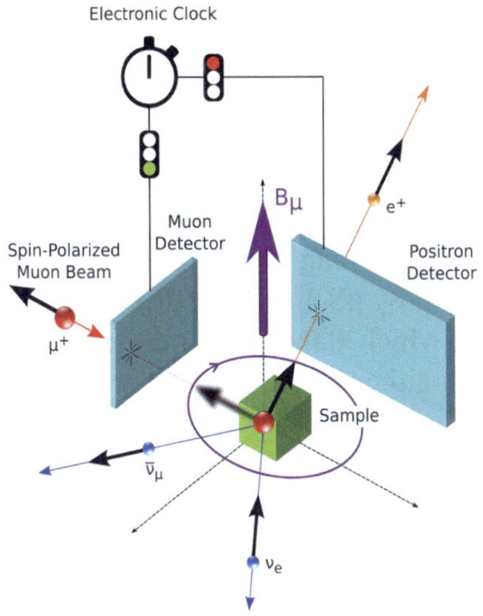

Figure 6.7. Muon spin spectroscopy.

Source: Junginger, T., Laxdal, R., MacFarlane, W. A., & Suter, A. (2024). SRF material research using muon spin rotation and beta-detected nuclear magnetic resonance. *Frontiers in Electronic Materials, 4*. https://doi.org/10.3389/femat.2024.1346235, Creative Commons License.

Since atomic muonium is about nine times lighter than hydrogen, there are isotopic effects for the dynamical properties that are connected to the difference in their total masses, even though the primary electronic characteristics of the two atoms should be almost identical. The significant elemental semiconductors, Ge and diamond, as well as GaAs, one of the well-known III-V compounds, exhibit the fundamental behavior for silicon. A significant advancement was made when a muonium state with a hyperfine interaction that only amounted to about 10^{-4} of the vacuum value was discovered in CdS, an II–VI compound(Böttcher and Dziesiaty, 1972). Because most approaches require rather high concentrations for detection, isolated hydrogen is a highly difficult species to study experimentally. The most prominent examples are the hyperfine interaction in Si, ZnO, and TiO_2, which has been measured for hydrogen using Electron Spin Resonance (ESR), and the donor and energy levels in silicon, which have been the subject of especially rigorous investigation using macroscopic methods. By measuring the hyperfine interaction and probing the electronic spin transitions when a magnetic field is applied, the electron spin resonance technique may assess paramagnetic defects in minerals at the microscopic level. Electron-Nucleus DOuble Resonance (ENDOR) is a supplementary technology that uses suitable stimulation with electromagnetic radiation to directly examine the spin changes of the interacting nucleus. The characterization of hydrogen defects is restricted to a few pertinent cases, namely Si, ZnO, the hyperfine interaction obtained for muonium with μSR, because these measurements require that the defect under investigation be relatively abundant. After accounting for the ratio of the muon magnetic moment (μ_{μ}) to that of the proton (μN)($\mu_{\mu}/\mu_{N} = 3.183$), the results for hydrogen utilizing ESR techniques are very similar. However, these values are around one order of magnitude different for ZnO, indicating that the configuration being probed by μSR and ESR/ENDOR is not the same. The first law of thermodynamics, which states that energy can be transformed from one form to another, but can neither be created nor destroyed, must be kept in mind in this situation. Therefore, we are typically dealing with energy conversion that takes place through various dynamic processes for all kinds of energy devices as well as on the atomic scale inside an energy material. Rechargeable batteries and fuel cells are two examples of energy-conversion devices. Both are electrochemical devices that effectively and immediately convert chemical energy into electrical energy. Both devices have a very similar general design and operation. They are made up of a cathode/electrolyte/anode stack, where energy conversion takes place by the movement of charged ions through the electrolyte and an accompanying flow of electrons in the outer circuit. A new generation of energy materials must be created in order to enhance these device functionality, longevity, safety, and sustainability. Understanding, managing, and improving ion diffusion within and between the various materials and components is one of the primary challenges in this case. Developing and applying cutting-edge experimental methods to investigate ion dynamics within the materials on the atomic length scale is essential to bringing about a paradigm change in this discipline. Energy-related materials and devices have seen significant advancements and performance enhancements in recent decades. This is most obvious to nonspecialists in the context of rechargeable batteries, which we use on a daily basis in devices like computers, smartphones, electric automobiles, etc. Until recently, electrochemical techniques such as electrochemical impedance spectroscopy (EIS) or potentiostatic/galvanostatic intermittent titration techniques (PITT/GITT) were used. Although the effectiveness of such methods cannot be disputed, it is crucial to realize that they do not directly produce intrinsic material qualities from the perspective of pure materials science. The explanation for this is because these methods necessitate building a functional apparatus, which may be a complete battery cell or a portion.

CHAPTER 7

X-RAY PHOTOELECTRON SPECTROSCOPY

7.1. INTRODUCTION

There are five senses that we use to interact with our environment. For the first three, signals must be transmitted via an interface of some kind, such as our skin, taste buds, or smell receptors. Two different types of matter that come into direct contact with one another are represented by an interface. These could be in the same or distinct phases, i.e., solid, liquid, or gas. The physical characteristics of the layers in contact determine how these different types of matter interact. The elements that are present and the ways in which these elements link to one another define the physical characteristics of matter in one way or another. Speciation is the term used for the latter. The difference between aluminum in its metal form and aluminum in its oxide form (Al_2O_3) is an example of speciation. Aluminum has a wide range of characteristics and occurs in two distinct oxidation states (Al^0, Al^{3+}) (Nestler et al., 2019). The latter is inert, whereas the former can be extremely explosive when the powder form is in an oxidizing environment; when combined with ammonium perchlorate, it served as a booster rocket propellant for the space shuttle. The popular commodity, aluminum foil, is made of metal. However, because it is covered in a thin oxide layer that spontaneously recovers if compromised, this is totally inert to the environment. Another name for this layer is a passivating oxide. There is no naturally occurring aluminum metal. Over the past 200 years, the cost of manufacturing this commodity has drastically dropped. It is said that Napoleon III gave his favorite guests aluminum cutlery as a sign of respect, while the less favored guests received gold cutlery. The aluminum alloy was formerly thought to be more valuable than gold. Similar to aluminum foil, the majority of materials that exist in the liquid or solid state have a surface layer that differs from the material underneath. This difference may be structural, chemical, or both. The shape of the exterior layer determines how a material is viewed. The bulk refers to the underlying material. The higher permeability of gases, which results from a lack of intermolecular interactions and the velocity of the constituents in air, N_2 and O_2 travel on average of 500 m/s, with any subsequent collisions defining pressure, is another reason why gases are not taken into consideration.

Two categories can be used to explain why a solid or liquid surface's physical characteristics may differ from the underlying bulk: External Forces, i.e., Adsorption and/or Corrosion of the outer surface. Examples of materials with a stable oxide (passivation layer) that is only a few atomic layers thick (1 nm) on the outside include silicon and aluminum fragments. Note that air is a reactive medium. Since both water vapor and CO_2 are found in air, it is true that water vapor catalyzes the adsorption of CO_2 on a variety of metallic surfaces. Internal Forces, or forces transmitted by surface free energy, can result in structural alteration when any long-range atomic structure abruptly ends. This might affect the outermost atomic levels. The following are a few physical characteristics that may be impacted by the changes like adhesion, adsorption, biocompatibility, corrosion, desorption, and electrical characteristics of the interface, reactivity, including heterogeneous catalysis, texture, and so on. It will be easier to comprehend how the corresponding solid or liquid interacts with its environment if the surface composition and speciation can be described. It is possible to alter or customize these attributes as needed. This has led to a number of technological advances in practically every field where surfaces are involved. The following are some fields where these changes have been used: adhesion research, automotive industry, biosciences, electronics, medical sector, and pharmaceutical industry. A lot of these innovations have evolved as a result of creating new materials that were not found in nature before or by customizing particular surface characteristics. Similar to aluminum foil, these are all manufactured; examples include the polymers and the production of superconducting oxides. The surface of a solid or liquid can be described in a number of ways. The simple definition is that an object's surface is its outermost or highest boundary. However, the term boundary loses its meaning at the atomic level since bound electron orbits are extremely diffuse. A different way to define a surface would be as the area that determines how a solid or liquid interacts with its environment. According to this concept, a surface can cover hundreds of atomic layers (100 nm or more) or as little as one atomic layer (0.1–0.3 nm), depending on the material, its surroundings, and the desired attribute.

7.2. RESOLUTION

Think of a human hair strand to put these dimensions into context. Its diameter ranges from 50 to 100 µm. The outer surface atoms have a diameter of about 0.2 nm. Since the spatial resolution is diffraction limited to values less than 1 µm, this cannot be seen even under the specialized optical microscope. Only very few methods can achieve the required magnification, with transmission electron microscopy (TEM)(Figure 7.1)being the most often used. However, because TEM is a microscope, it can only show the physical structure.

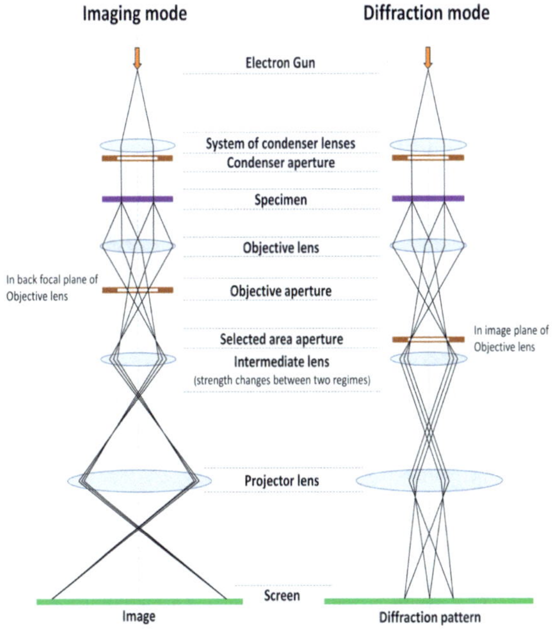

Figure 7.1. Transmission electron microscopy.

Source: https://upload.wikimedia.org/wikipedia/commons/thumb/a/a0/Schematic_view_of_imaging_and_diffraction_modes_in_TEM.tif/lossless-page1-1024px-Schematic_view_of_imaging_and_diffraction_modes_in_TEM.tif.png, Creative Commons License.

Spectroscopy or spectrometry is needed to know the chemistry. Few spectroscopies can provide the chemistry active over a solid outermost surface, or within its outermost 10 nm, but there are many of them. X-ray photoelectron spectroscopy (XPS) or electron spectroscopy for chemical analysis (ESCA), is the widely used of the few available techniques(Barr, 1991). When determining the elemental composition of a solid outer surface, XPS, also known as ESCA, is the widely used electron spectroscopy method. When Kai Siegbahn discovered that speciation could be inferred only from the photoelectron and Auger electron emissions, he first proposed the abbreviation ESCA. Because it can identify and measure the elemental composition of any solid surface that is 10 nm or less in thickness, with all elements from Li to U observable, XPS is popular. Because of their extremely low photoelectron cross sections and that XPS is designed to investigate core electrons, H and He are undetectable. This is based on the premise that the element of interest exists at >0.05 atomic %.

Auger electron spectroscopy (AES)(Figure 7.2)is the technique closely related to XPS in that it exhibits a similar surface specificity while being sensitive to the same elements (Li–U), with the exception of ultraviolet photoelectron spectroscopy (UPS), which can be considered an extension of XPS since it measures the valence band photoelectrons.

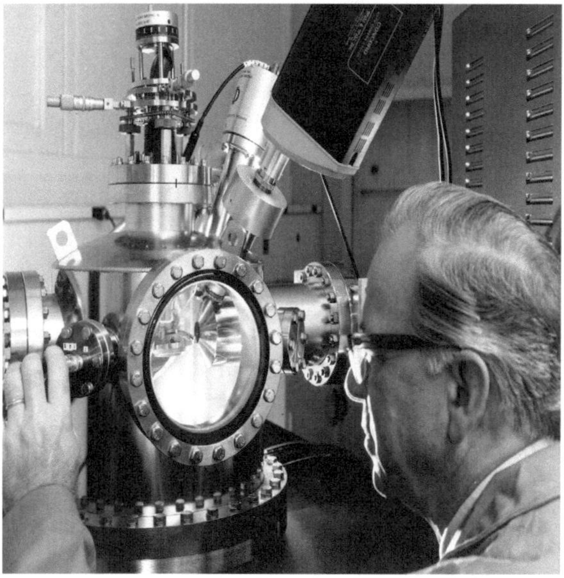

Figure 7.2. Auger electron spectroscopy (AES).

Source:https://upload.wikimedia.org/wikipedia/commons/3/3f/HD.6C.037_%2811856519893%29.jpg, Creative Commons License.

Although it comes at the expense of sensitivity, its strength is its increased spatial resolution. The elemental composition of solids can also be effectively determined using energy-dispersive X-ray analysis (EDS or EDX) and wavelength-dispersive X-ray analysis (WDX). They are common than XPS when paired with scanning electron microscopy (SEM), (Figure 7.3)and the acronym electron probe microanalysis (EPMA) is frequently employed. However, as they offer average elemental concentrations spanning a depth of at least 1 μm below the surface, these are not regarded as accurate surface analytical techniques. This depth differential is significant because the chemical composition, speciation, and/or electronic structure found over the outer few atomic layers of the corresponding solid determine all of the surface chemistry that occurs between various forms of matter.

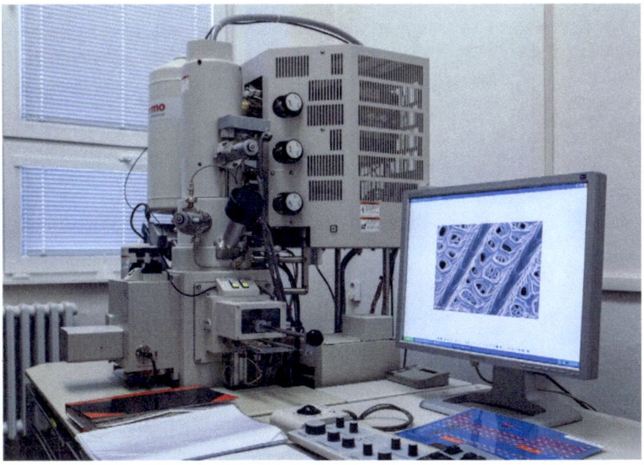

Figure 7.3. Scanning electron microscopy.

Source: https://upload.wikimedia.org/wikipedia/commons/f/f1/Scanning_electron_microscope_-_UFCH_JH_%282020%29_01.jpg, Creative Commons License.

These can be different from those observed at least one nanometer below the surface. Three to four atomic layers are represented by a surface coating of 1 nm. EDX analysis of aluminum foil shows spectra that are largely dominated by aluminum peaks. Heinrich Hertz discovered that electrically isolated metallic objects kept in vacuum had a better capacity to ignite when exposed to light in the 1880s, which is when XPS first emerged. This phenomenon, known as the Hertz effect, also made it possible to determine the work function (ϕ) of the corresponding metal object and the ratio of Plank's constant over electronic charge (h/e) when adjusting the energy of the irradiation source. Albert Einstein described this effect in 1905 as the result of photons transferring energy to electrons that are confined within the atoms that make up the corresponding metallic objects. He demonstrated that if the energy transfer exceeded the energy that bonds the electron to the corresponding metal atom or solid, this caused electron emission from metallic objects. Einstein received the Nobel

Prize in Physics in 1921 for this and for introducing the idea of the photon, which is a packet of energy with zero rest mass. The net positive charge that accumulates on photoelectron and Auger electron emission from electrically isolated objects might be interpreted as the cause(Cazaux, 2010).

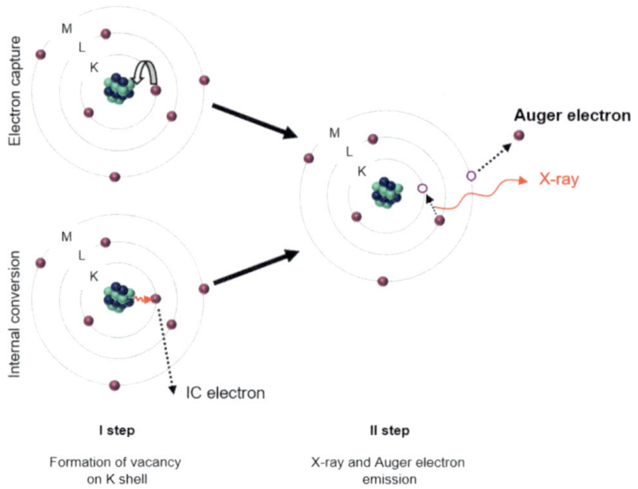

Figure 7.4. Auger electron emission.

Source: Gharibkandi, N. A., Gierałtowska, J., Wawrowicz, K., &Bilewicz, A. (2022). Nanostructures as radionuclide carriers in auger electron therapy. *Materials, 15*(3), Article 3. https://doi.org/10.3390/ma15031143, Creative Commons License.

7.3. AUGER EMISSION

Auger electron emission (Figure 7.4)or fluorescence also occurs in conjunction with photoelectron emission. While fluorescence was called by George Stokes, who provided a physical explanation for this phenomenon, Auger emission was originally recorded by Lisa Meitner and is named after Pierre Auger. However, it was not until Kai Siegbahn and his colleagues built a device that could analyze core photoelectron emissions to a high energy resolution to enable speciation analysis that the full potential of XPS was acknowledged. Kai Siegbahn received the 1981 Nobel Prize in Physics in recognition of this. After Siegbahn's initial achievement, a series of investigations and tools led to a solid knowledge of XPS. The main cause of this relatively recent development is the impossibility to achieve the vacuum conditions needed for the analysis of such surface regions. This application is typically restricted to the investigation of solid surfaces due to its demand for vacuum. It is possible to analyze gases and liquids, but only with the use of extremely specialized equipment and/or sample preparation techniques. Photoelectron emission refers to a one-step process where a photon ejects an electron that was first attached to an atom or ion. Photons are annihilated during photon–electron interaction with full energy transfer because they are a massless, chargeless packet of energy. Should this energy be adequate, the electron will be released from both the solid and the atom/ion. The amount or quantity measured is the kinetic energy (K.E.) that is still present on the released electron. This is helpful since it is distinct in nature and depends on the electron binding energy (B.E.), which varies depending on the element and environment. Photon emission is the outcome of the other step, which is known as fluorescence.

These are explained in terms of X-ray notation. Assuming that the element of interest exists at >0.05 atomic percent (%), two parameters characterize XPS capacity to detect and quantify the elemental composition and speciation present over the outer 10 nm or less of any solid surface. They are surface specificity, or the capacity to distinguish the signal from the surface area in comparison to that of the underlying region; and sensitivity, or the capacity to detect the signal of interest given the restriction of the reduced volume from which the signal emanates. Surface specificity results from an electron's short flight path within a solid before losing a portion of its energy; typically, this is less than 10 nm. X-rays have the ability to reach micrometers below the surface. The signal will disappear into the spectral background if energy is lost. Nearly all photoelectrons generated from atoms or ions located more than roughly 10 nm below the surface exhibit this behavior.

The remaining distinct signals which produce the spectral peaks that are seen, come only from the surface region. An adsorbed surface layer of a certain thickness will therefore serve to partially block all signals coming from the substrate underneath. This lessens XPS sensitivity to these components then. Sensitivity is determined by the spectral background level and photoelectron cross section, respectively(Powell and Jablonski, 2009). According to the photon energy, the yield of electrons generated is described by the photoelectron cross section. The main cause of H and He non-detection in solids is their low photoelectron cross sections when exposed to X-ray light. Since the ambient pressure, or more specifically the vacuum in which the analysis is performed, regulates the density of molecules in the gas phase and, consequently, the flight path of any photoelectrons emitted from the surface, it can also have an impact on sensitivity. This limits the flow of electrons from the sample to the detector. If the pressure used to perform the analysis is too high, contaminant overlays may also develop during the analysis timeframe. The success of the majority of advanced analytical techniques is determined by two factors: the instrumentation and the expertise and experience of the individual interpreting the generated spectra. The environment inside this device must be such that no external magnetic or electrostatic fields can influence the photoelectron and Auger electron emissions.

7.4. PRESSURE

The area between the sample and the detector can be traversed by the photoelectron and Auger electron emissions. Throughout the analysis process, the sample from which the emission originates cannot be altered in any way. The pressure in the area of interest defines a vacuum. The pascal (Pa) is the SI unit for pressure. The Torr is the widely used unit in vacuum science and technology. Another unit is the millibar (mbar). The Langmuir (L) is another term. This is the pressure-time integral needed to create a monolayer; assuming a surface with a unit sticking coefficient, 1 L is equivalent to around 10^{-6} Torrs. Gathering spectra from a metal surface is likely the best way to demonstrate this. There is still a clear C-1s peak even on surfaces that have been previously cleaned. Depending on the sample and sampling volume, quantifying this signal yields levels ranging from 1 to 15 atomic %. This signal results from gas-phase molecule adsorption, which takes place even in UHV environments. The first step in comprehending how this is possible is realizing that, in any gas at room temperature and atmospheric pressure, there are approximately 2×10^{19} molecules traveling in random directions at velocities exceeding 100 m/s, with the kinetic theory of gases providing the description of the average velocity. These molecules have a limited chance of colliding with a solid surface, such as the sample or the walls of the container, because they are mobile. The measured pressure is the result of more than 10^{23} collisions/cm^2·s in atmospheric conditions. Despite the invention of the vacuum pump in 1650, it took nearly three centuries to achieve vacuum conditions superior to $\sim 10^{-7}$ Torr.

Reducing the number of molecules in a small area as comparison to its surroundings creates a vacuum. Pressure is created by the existence of these molecules or, more specifically, by their collisions with their environment. However, a highly specialized pumping system and a chamber with exceptionally low outgassing properties are needed to obtain UHV conditions. The removal of surface adsorbates produced when any surface is exposed to ambient gases is known as outgassing. Stainless steel with a mu-metal liner or mu-metal itself is used to make vacuum chambers. These substances are employed because of their low rate of outgassing, low rate of corrosion, low vapor pressure due to the high melting point of stainless steel, structural integrity in both thermal and ambient environments, relative simplicity of manufacture (machining) and cost-effectiveness. There is less usage of high vapor pressure materials such as rubber O-rings, epoxy resins, borosilicate glass, and adhesive tapes. Chambers that can withstand UHV conditions are typically made in spherical or cylindrical shapes to further improve the structural integrity. After that, access is made possible by ports that are joined utilizing specific welding techniques. After that, units are joined using flange/gasket combinations that have been specially constructed. The Conflat system is the sort of flange/gasket that is frequently utilized in UHV systems(Peacock, 1998). The flanges press into a copper gasket when squeezed, creating a tight UHV seal. However, these gaskets need to be changed after every use. Al, Au, and In are soft gasket materials that have been utilized to create UHV conditions. Carbon-based O-rings are frequently utilized for high voltage applications since they are reusable and typically more affordable. An attached self-contained chamber that can be closed by a number of atmosphere and/or analysis chamber valves is used to introduce samples from the atmosphere. Depending on the sample outgassing rates, this enables the sample to be pushed down from the atmosphere to a pressure of roughly 10^{-7} Torr in 10^{-20} minutes. To enable sample transmission, the valve between the introduction and analysis chambers is opened once the pump has been lowered.

7.5. PUMPS

Multiple pumps, valves, plumbing tubing, and vacuum gauges to record the vacuum generated make up the plumbing utilized in XPS equipment. The best pumps for supporting momentum transfer pumps and evacuating from atmospheric pressure to about 10^{-3} Torr are positive displacement pumps. For chambers that have already been pre-evacuated to within the 10^{-3} Torr range, momentum transfer pumps can provide a vacuum as low as 10^{-10} Torr. Turbomolecular pumps are the widely utilized momentum transfer pumps in XPS devices. In order to empty an introduction chamber, these are usually set up in conjunction with a positive displacement pump. The ideal pressure that can be attained in a reasonable amount of time is in the 10^{-7} Torr range since an introduction chamber is frequently exposed to the atmosphere. However, once the valve between the two is opened, this is adequate to for a sample to be moved into the analysis chamber held under UHV; in other words, it prevents the analysis chamber from becoming contaminated. Entrapment pumps may generate a vacuum as low as 10^{-10} Torr and are self-contained devices that do not need any additional pumps. Entrapment pumps work by either ionizing gas-phase molecules into a solid state or by trapping or condensing them into a solid state. The prevalent ones in XPS equipment are ion pumps, to which titanium sublimation pumps are frequently added. To maintain UHV conditions, ion pumps are run constantly. With sublimation pumps, however, vacuum conditions can be improved by operating them intermittently. The initial pump down of these chambers is accomplished using a combination of displacement and momentum transfer pumps because neither should be used at pressures higher than approximately 10^{-5} Torr. When building a chamber that can withstand UHV conditions, pumping speed must also be taken into account. The longest pump down periods is from VHV to UHV. This occurs because the pressure is determined by the rate at which surface adsorbates are removed. Any metal surface exposed to airborne gases will produce adsorbates. Regardless of the pumping mechanism employed, pumping a vacuum chamber from the atmosphere to UHV under ambient circumstances can take weeks. This also explains why samples are not moved from the atmosphere into the analysis rooms. These chambers are heated anytime they are exposed to the environment in order to hasten the removal of surface adsorbates. This means that the chamber must be heated to 150–200°C while being pumped for 12–48 hours. All temperature-sensitive components and connections need to be unplugged in order to accomplish this.

Figure 7.5. X-ray tube.

Source: https://upload.wikimedia.org/wikipedia/commons/7/72/Hvtubes.jpg, Creative Commons License.

7.6. X-RAYS

By aiming an electron beam at a metallic material with enough energy, X-ray tubes (Figure 7.5)generate X-rays. The cathode is the source of electrons, and this metallic object is known as the X-ray anode. Cathodes are a thermionic source. Al is widely utilized solid in XPS, even though any solid can theoretically be used as an X-ray anode. This popularity results from Al-Kα X-rays with comparatively high energy and intensity. Al-Kα X-rays have minimal energy spread. The efficiency of Al as a heat conductor. These anode robustness and ease of production is another reason. Kα X-rays are the X-ray emissions generated in the case of Al anodes, which contributes to the high intensity. Although Kα1 and Kα2 X-rays are essentially part of Kα X-rays, they are sometimes referred to as $K\alpha_{1,2}$ X-rays or, simply, Kα X-rays because of their small energy separation(Goldstein et al., 2018).

When an incoming energetic electron scatters inelastically, an Al-1s electron is released. An electron from a 2p level subsequently fills the Al-1s core hole that is created. Quartz single crystals are used in monochromatic sources on commercially available XPS instruments because they are UHV compatible, that means they do not outgas or degrade, and they are relatively inert under atmospheric conditions. These sources produce X-ray spots with diameters ranging from roughly 10 μm to several thousand micrometers. When compared to conventional sources, the primary advantages are their capacity to generate a concentrated X-ray beam without the source needing to be close to the sample and the generated X-ray lower energy spread. The main drawbacks in contrast to conventional sources include their increased expense and the finite quantity of anodes that can be used in a given design in order to meet the Bragg criterion. When examining insulating surfaces, there is a higher chance of sample charging because of the higher photon flux per unit area compared to normal sources.

These sources are examined despite the fact that they cannot be used to produce core-level photoelectron emissions because they offer valence band spectra with higher sensitivity and better energy resolution than can be obtained with monochromated sources. Nonsynchrotron-based XPS equipment that have standard or monochromatic X-ray sources are frequently equipped with these. The X-ray energy and the valence band energy match closely, which results in increased sensitivity. The narrower energy distribution of the generated photons results in the enhanced energy resolution. He gas is passed through the discharge region, which is created by applying a potential difference of about 1 kV between the anode and cathode regions, in gas discharge lamps used in UPS. He electrons are excited to higher stationary states as a result. Photons with an energy equal to the difference between the states accessible are emitted during de-excitation. The widely used primary emission in UPS is at 21.2 eV. A secondary, weaker emission at 40.8 eV is also reached. These sources use magnetic fields to push electrons moving at nearly relativistic speeds into curved pathways, which creates photons. The intensity of the photons, which are released tangential to the electron beam, is inversely proportional to the electron energy cube and proportional to the radius of curvature of the electron beam. The photon energy is constant between about 1 eV and about 10 keV. The benefits of such sources include enabling the selection of any photon energy, which allows for the examination of resonance effects related to ionization thresholds and offering a beam with a narrower energy spread, which enables better overall energy resolution with a value of ~0.1 eV or higher. Producing a photon flux per unit area that is at least 100 times more than what is achievable with conventional or monochromated sources is another benefit. Using zone plates to create a highly collimated beam that can be focused to about 0.1 μm, which enables the highest possible spatial resolution is another advantage. Permitting interaction between the sample and a polarizable beam can be assumed as a benefit. Drawbacks of Synchrotron sources are: synchrotron is required; while these are accessible at different research organizations, their time is limited and might be expensive, because these sources create more flux per unit, there may be higher sample deterioration during analysis, especially for organic samples. However, these are the preferred sources at research facilities exploring the frontiers of condensed matter physics, etc., due to their larger flux per unit area, pre-selectable energy range, and superior energy resolution for researching core-level emissions.

7.7. CHARGE MODERATION

In order to charge adjustment, electron sources are utilized in XPS as the cathode inside X-ray tubes and as a separate unit for directing electrons to insulating materials. While there are many types of electron sources, thermionic sources are the frequently found in XPS devices that are sold commercially. By heating a substance until the bound electrons have sufficient energy to surpass the material work function (ϕ), thermionic sources produce electrons. When this occurs, they escape into the vacuum, where the fields can be used to steer them to the sample at the appropriate energy (10 eV–10 keV). As beam energy decreases, beam diameters for all sources grow due to space charge effects. LaB_6 crystals (ϕ value of 2.6 eV) and W filaments (ϕ value of 4.5 eV) are the two main thermionic sources in use. LaB_6 sources operate at lower temperatures (2,000 K–LaB_6 / 2,700 K–W) and produce higher currents because to the lower ϕ(Boulos et al., 2023). The lifetime is increased by the decreased energy spread caused by this lower working temperature. Focusing to smaller beam diameters is also made possible by a reduced energy spread; with LaB_6 sources, spot sizes of less than 0.1 μm at 10 keV are achievable, compared to a few micrometers for W filaments. However, LaB_6 sources are more expensive and more sensitive to vacuum conditions. XPS uses ion sources to help with charge adjustment when examining insulating materials or to enable sputtering of the surface of interest. Even though there are many different types of ion sources, EI sources are frequently found in XPS devices that are sold commercially. Inert gas ion electron impact (EI) sources have been the widely used method for XPS sputtering. These sources include an area where electrons with an energy of around 70 eV from a heated filament introduce and irradiate the gas of interest (Ar). This causes the corresponding gas atoms or molecules to emit electrons, creating the corresponding positive ions. After that, the ions are removed via a plate aperture that is kept at a potential that is negative in relation to the ionization zone. A Wien filter and a minimal bend are inserted into the ion column to guarantee the purity of this beam. Wien filters are made up of crossed magnetic and electrostatic fields that deflect the ion beam in accordance with its charge, mass, and energy. This efficiently eliminates any multiply-charged ions from the beam as well as any undesirable neutral components. Due to their different focusing characteristics, both are unsuitable. A concentrated ion beam is then guided to the target spot on the sample by electrostatic lenses and deflectors. The deflectors enable rastering, or beam scanning across predetermined regions. To guarantee that an effective depth resolution can be achieved, these are set significantly larger than that which the X-ray beam analyzes. These sources may produce beams of medium brightness between 10 eV and 20 keV and currents up to 500 nA. The replaceable filament controls the lifetime of these dependable sources. The inability to focus the beam to spot sizes smaller than roughly 10 μm is the only disadvantage of XPS. Sputtering of organic layers using different cluster ion beams has garnered attention recently since it significantly reduces sample damage as compared to inert gas ion beams. Inert gas cluster ions and organic cluster ions are the two types of cluster ions of interest. The forms of organic cluster ions include C_{60}^{+}, C_{84}^{+}, $C_{24}H_{12}^{+}$or coronene, and others. These employ sublimation ovens to generate the necessary gas, which is ionized by electron impact after the powder is heated to about 170°C to form coronene gas. The filament and the source material control the lifetime of these sources, which may produce low-brightness beams up to 20 keV to currents approaching 50 nA. The disadvantages are that they are more expensive than inert gas ion sources and that they cannot focus a cluster ion beam to spot sizes smaller than a few hundred micrometers. Each singly charged ion in an inert gas cluster consists of thousands of Ar atoms. These are created in a gas cluster ion beam (GCIB) source by cooling individual gas atoms in a supersonic jet, which condenses them into neutral clusters. After that, they are ionized by electron impact. An energy filter with both a high-energy resolution and a high transmission is necessary for efficient speciation analysis since XPS information is obtained from the $K.E._{XPS}$ of the electron emissions. While the latter enables sensitivity to be maximized, the former optimizes speciation identification by separating closely spaced peaks.

7.8. CYLINDRICAL MIRROR ANALYZER (CMA) AND CONCENTRIC HEMISPHERE ANALYZER(CHA)

Cylindrical Mirror Analyzer (CMA)(Figure 7.6)is a conventional multi-technique instrument and AES instrument. Concentric Hemisphere Analyzer (CHA), (Figure 7.7)spherical deflection analyzer (SDA),or electrostatic hemispherical analyzer (EHA), or hemisphere sector analyzer (HSA) is used in XPS(Verma, 2007). The ability to achieve better energy resolution, values greater than 0.01 eV is the main benefit of a CHA over a CMA.

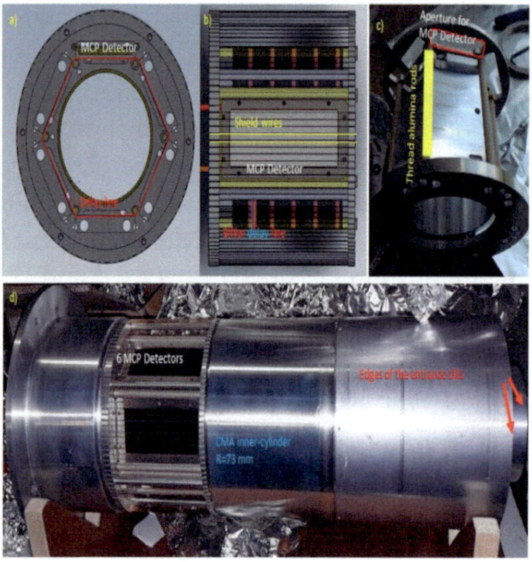

Figure 7.6. Cylindrical mirror analyzer (a-d).

Source: Penent, F., Cubaynes, D., Lablanquie, P., Palaudoux, J., Guilbaud, S., Moustier, O., Guigand, J., &Bizau, J.-M. (2020). Modification of a cylindrical mirror analyzer for high efficiency photoelectron spectroscopy on ion beams. *Atoms, 8*(4), Article 4. https://doi.org/10.3390/atoms8040063, Creative Commons License.

A CHA energy resolution is superior to the energy spread of an X-ray source. The X-ray source, not the CHA, determines the energy resolution in XPS. CHA may analyze higher regions than is allowed when using a CMA and shows a weaker K.E. dependence on the distance between the analyzer and the sample than a CMA. However, a CHA transmission is worse than a CMA because of its smaller collection solid cone. Transfer lenses make the surface of the sample more accessible. At this stage, it is also important to discuss a particular kind of energy analyzer known as the spherical mirror analyzer (SMA). A CHA and this energy analyzer, which are both employed in parallel imaging, use 180° hemispheres for electron pass. The distinction, however, is that the radii of the required electron paths are substantially less than a SMA inner sphere radius. The inner sphere has openings to provide access. There are two ways to obtain an energy spectrum in XPS.

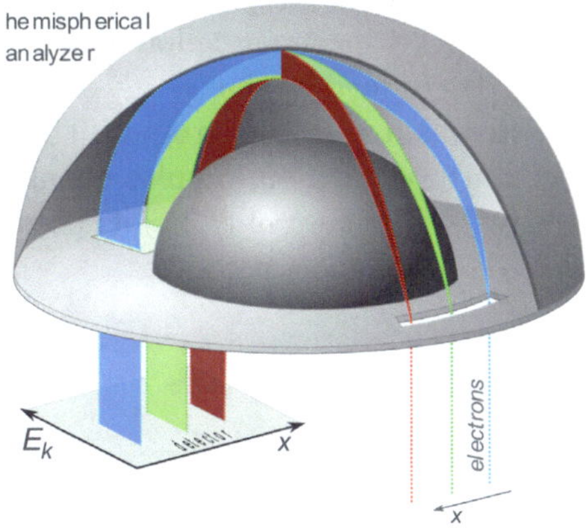

Figure 7.7. Concentric hemisphere analyzer.

Source: https://upload.wikimedia.org/wikipedia/commons/0/0c/Hemispherical_electron_energy_analyzer_working_principle.svg, Creative Commons License.

These are accomplished by scanning the corresponding CMA/CHA inner and outer cylinder/hemisphere potentials over the E_o range. This is known as the fixed retard ratio (FRR) mode or the constant retard ratio (CRR). Before entering the CMA/CHA, the electrons must be accelerated or decelerated to a constant E_o value. This is known as the fixed analyzer transmission (FAT) mode or constant analyzer energy (CAE). This can be used in instruments that use double-pass CMAs, but it is frequently used in instruments that use CHAs with the acceleration/deceleration carried out in the transfer lens. Potentials applied to the CHA hemispheres or the second stage of a double-pass CMA are maintained constant in both scenarios. The energy resolution in the CRR/FRR mode drops as E_o and, consequently, K.E. increase. Consequently, a percentage of $K.E._{XPS}$ is used to represent the energy resolution. The resolution of a high-quality CMA can be close to 0.1%. In the case of 1000-eV electrons, this corresponds to 1 eV. The energy resolution in the CAE/FAT mode can be represented in units of E_o since it is independent of $K.E._{XPS}$. This is called the pass energy, and lower values mean better resolution at the expense of sensitivity. High-energy resolution experiments employing monochromatic Al-Kα sources and a CHA conducted at pass energies of about 5 eV can therefore achieve an energy resolution of about 0.4 eV in the advanced equipment available. In order to achieve this resolution for 1000-eV electrons in the CRR/FRR mode, a CHA with a radius more than one meter would be needed. Measuring the energy of electron emissions and the quantity of electrons generated are both crucial in XPS. It is true that XPS spectra are plotted in energy versus intensity units, where the energy is determined by the energy analyzer and the intensity is determined by the detector recorded electron count. The detector must be able to record individual electrons, or function in pulse counting mode, in order to achieve the highest sensitivity. The current (A) units used to record this signal are subsequently converted to counts per second.

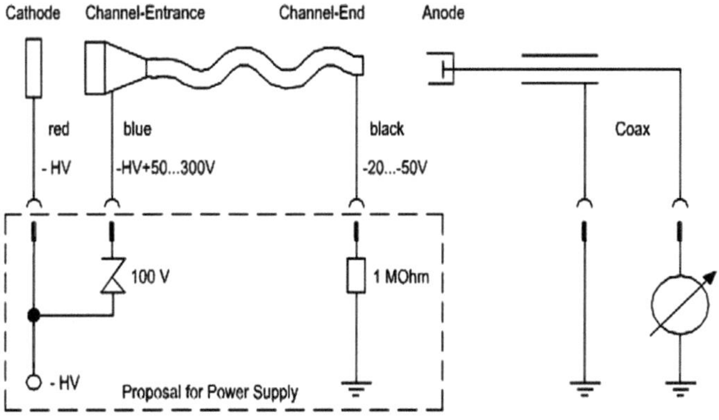

Figure 7.8. Channeltron.

Source: https://upload.wikimedia.org/wikipedia/commons/a/ab/Sch%C3%A9ma.jpg, Creative Commons License.

The flow of electrons, or charge, over a fixed location in space in a unit of time (seconds) is represented by the current. One electron can carry a charge of -1.602×10^{-19} C. Therefore, 1.602×10^{-19} A is the passage of this single electron per unit second. However, the 10^{-15} A range, which corresponds to the passage of 6241 electrons per second is where the sensitivity power of conventional pulse counting devices is found. Therefore, the detector must have a gain of at least 10^4 in order to record individual electrons. The electron multiplier (EM) can function in the UHV conditions needed by XPS. EM refers to a class of detectors that work by intensifying the signal produced when an electron strikes the EM initial or outermost surface, or the surface that is exposed to the incoming electron beam. Among these are channeltrons, also known as channel electron multipliers (CEM), discrete dynode Ems, and detectors using microchannel plates (MCP)(Gys, 2015). A metallic conversion electrode and several electrically separated acceleration electrodes make up a discrete dynode EM. The term dynodes refer to these electrodes. When the incoming electron hits the conversion dynode, which covers an area of around 10 × 10 mm, several secondary electrons are produced. After that, the secondary electrons are passed into the first acceleration electrode, which is the next nearest electrode. This electrode then generates more secondary electrons, which are accelerated into the second acceleration electrode, which is the next acceleration dynode, and so on. These have a maximum gain of 10^9. However, when the incoming electron count rate surpasses approximately 1×10^6 cps, these detectors begin to saturate. The linearity of the input to output signal is lost as a result.

7.9. CHANNELTRON

A horn-shaped continuous dynode device coated on the inside with an electron-emissive substance, such PbO, is called a channeltron, or CEM(Figure 7.8). Similar to a discrete dynode EM, an electron impacting the channeltron entrance causes an electron cascade. The potential difference between the two ends accelerates the electrons into the horn. Channeltrons can achieve gains about 10^8 while sharing many traits with discrete dynode EMs, such as a similar linear range. The main benefit of these detectors over XPS is that the manufacturer can design the opening such that, for instance, an array of around 5 × 15 mm detectors can be arranged side by side along a particular plane of interest (or the energy-dispersive axis of an energy analyzer). With the help of this 1-D array, data may be collected concurrently over a range of energies. The data can then be plotted as a function of energy, or all of the data within a predetermined range of energies can be put together to increase sensitivity overall. Since an MCP is made up of a 2-D array of parallel glass capillaries with an inner diameter ranging from 10 to 25 μm and an axis that is around 7° off the trajectory of the signal to be recorded, it can be viewed as an extension of a 1-D channeltron array. These capillaries, like channeltrons, have an inside coating of an electron-emissive substance that permits ions or electrons to strike or hit the inner wall and produce an electron cascade. The gain from this cascade can reach 10^4. A gain of 10^6 can be achieved by using two channel plates with one 180° turned relative to the other (Chevron MCPs). However, when the incoming electron count rate surpasses approximately 3×10^5 cps, these detectors begin to saturate. One of three methods can then be used to collect the generated current pulse:

an anode collector, a phosphor screen, or a delay line detector (DLD). A metal electrode coupled to extremely sensitive pulse counting electronics is all that makes up an anode collector. When electrons strike or hit phosphor screens, photons are produced. Direct imaging of the signal of interest is made possible by the use of these in conjunction with an optical lens and camera configuration. These are known as charge-coupled devices (CCDs)(Figure 7.9).

Figure 7.9. Charge-coupled devices.

Source: https://upload.wikimedia.org/wikipedia/commons/9/94/ArgusCCD.jpg, Creative Commons License.

Further amplification is achieved so that the total gain approaches 10^8 because phosphor screens emit several photons for every electron. Similar to an anode encoder, the DLD records the output directly on a wire; the only distinction is that the location at which the signal is recorded along the wire permits the registration of the point at where the initial impact on the MCP occurred. The time it takes for the signal to move along the cable in both directions is used to determine this position. The signal is identified as originating from the middle of the MCP stack if the times are the same. Two sets of wires, each of which is a single wire flowing back and forth along a certain axis, are used to create a 2-D image. These are a relatively new technology that is quickly being used in XPS for both energy and spatial imaging as well as pulse counting. The amplification factor for EMs depends on the K.E. of the striking electrons; the higher the K.E., the greater the amplification. The type of EM and the operating conditions affect the amplification factors. It is advisable to employ the CAE/ FAT mode on the energy analyzer that is used to filter the electron emissions in order to do quantification. This is achieved because it guarantees that all of the electrons contacting the EM have energies within the same K.E. range. To improve EM sensitivity, post-acceleration of the electron emissions can be used. Imaging refers to the process of mapping electron emissions so that their distribution reflects the spatial distribution of elements and/or their speciation throughout the sample surface. Imaging has advanced significantly over the past few decades, and research is still continuing, due to advancements in data storage and sensor design. A number of strategies have been devised by the different XPS manufacturers. These methods can be separated based on whether the images are captured in a serial or parallel methods.The distribution of elements and/or their speciation in relation to depth is also interesting. Imaging of electron emissions at various takeoff angles is one technique used to obtain the data required to derive depth distributions, even if it is within the sampling depth. Creating maps in serial mode necessitates gathering information point-by-point while reassembling the image pixel by pixel in real time. This can be done in a number of ways, including scanning the sample stage, scanning the surface of interest using an X-ray probe, and controlling the region from which the electron emissions are collected.

7.10. IMAGES

Images are obtained by merely moving the sample stage while continually gathering the electron emissions. When images with a diameter of several hundred micrometers are of interest, the second method, also known as the probe defined mode, is employed. Since the crystal in these units concentrates the X-ray beam, this mode necessitates a monochromotic X-ray source. Rastering the electron beam that creates the X-ray beam across the X-ray anode surface causes scanning(Masuch et al., 2023). The electron raster pattern and the timing of the detector electronics are then synchronized to create the image. Since the X-ray beam diameter determines the resolution, this method can be used with devices that use a CHA or a CMA. The energy analyzer can be used in its wide-open mode i.e., lowest-energy resolution, which maximizes electron emission transmission and, thus, sensitivity. By scanning the energy filter during imaging, complete spectra can be obtained for every point (pixel). There are very efficient, commercially accessible devices that use this method to collect photos. When images over a region spanning several hundred micrometers in diameter are of relevance, the third method, also known as the lens defined mode, isemployed. Any X-ray source can be utilized in this mode since it regulates the region over which the transfer lens or area defining aperture collects and transfers the electron emissions to the detector. An image is created by scanning the region where the electron emissions are collected. The voltages on the deflector units inside the transfer lens, which is located before the area-defining aperture, are ramped to achieve this. This method is therefore restricted to instruments with a CHA, which can function in either the CAE/FAT or CRR/FRR modes, due to the requirement for the transfer lens/area defining aperture. A CHA and an array of detectors positioned along the CHA energy-dispersive axis offer the following potential outcomes: derive pixel-specific energy distributions with a limited data density and over a narrow energy span, adds information from nearby energy channels to raise the overall sensitivity. More recently, maps of the distribution of elements and/or their speciation across a surface have been created using animaging technique called parallel imaging. With a number of useful commercial devices, this is utilized to examine regions less than a few hundred micrometers in diameter. Without scanning the sample stage, the X-ray source, or the deflectors in the transfer lens, electron emissions from the full field of view are transmitted simultaneously. The electron emissions within a predetermined $K.E._{XPS}$ range are projected electrostatically and/or magnetically from the sample surface through a CHA or a SMA energy analyzer and onto a 2-D detector to achieve imaging in this mode. This calls for a 2-D MCP in addition to extra lenses positioned before and after the energy analyzer to guarantee that spatially scattered signals (electrons) enter and exit the energy analyzer along parallel pathways. A baffle inside the SMA, positioned at the focus point where the higher energy dispersion takes place, limits the energy of the electrons to be imaged in this arrangement. The energy of the electron emissions causes focal points to diverge in space, just like in photon optics. In order to preserve image resolution and lessen the impact of chromatic aberrations, this energy filtering must be done. The spherical aberrations are the elements that influences spatial resolution. By keeping the electrons near the instrument optical axis, they are reduced. The following are the advantages of this method over serial approaches: Spatial resolution improvements are observed. With this imaging method, values slightly better than 3 μm are achievable, acquisition times are shortened with the same sensitivity. It only takes a few seconds to collect 256 × 256-pixel photos. The fact that each pixel represents electron emissions at a certain energy is a significant drawback of the parallel imaging technique. The instrument can be returned to spectroscopic mode to gather spectra from localized regions of interest, which provides a more accurate understanding of the peak shape and background and, consequently, more accurate quantitative data, or more images can be taken away from the peak of interest, which provides an estimate of the background signal. There is also a type of parallel imaging that simultaneously relays the data required to derive depth distributions inside the sampling depth. This is possible because photoelectrons that leave the sample surface at various takeoff angles can be distributed along the plane normal to a CHA energy-dispersive plane and mapped on an MCP(Gunawardane and Arumainayagam,

2006). The depth from which the photoelectrons arise is determined by the takeoff angle.

7.11. SIGNAL INTENSITY

The ability to distinguish signals originating from physically different locations when displaying a distinct composition or specification is referred to as spatial resolution. When used in the parallel imaging mode, the commercially available XPS equipment using a monochromated Al-Kα source and a CHA or SMA can achieve a spatial resolution of < 3 μm. When employing a CHA with images taken in a serial method using the probe specified mode, this spatial resolution rises to values < 10 μm. When taking pictures in a serial method with the lens-defined mode, a further increase to about 15 μm is observed. Only alternative methods, such as those resulting from the development of energy-filtered photoelectron emission microscopy (PEEM)(Figure 7.10)or, X-ray photoelectron emission microscopy (XPEEM), (Figure 7.11)can currently be used to determine submicrometer scale spatial resolution values in XPS.

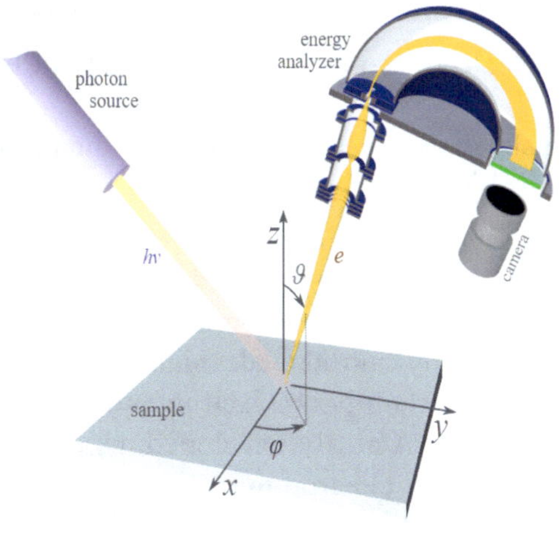

Figure 7.10. Photoelectron emission microscopy (PEEM).

Source: https://upload.wikimedia.org/wikipedia/commons/3/3f/ARPES_analyzer_cross_section.svg, Creative Commons License.

Employing a specially designed double CHA in conjunction with an Al-Kα monochromatic X-ray source has been shown to achieve a spatial resolution of about 0.5 μm. Using a synchrotron source further improves this spatial resolution to about 0.1 μm, and more improvements are anticipated.

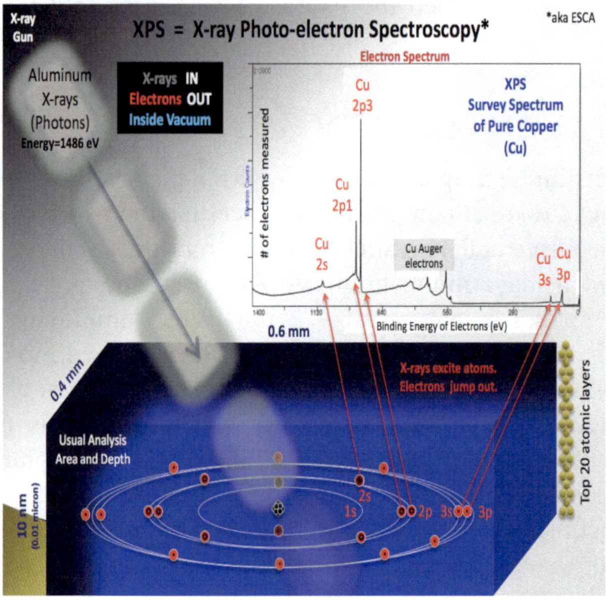

Figure 7.11. X-ray photoelectron emission microscopy (XPEEM).

Source: https://upload.wikimedia.org/wikipedia/commons/0/07/XPS_PHYSICS.png, Creative Commons License.

The highly specialized procedures PEEM and XPEEM analyze electron emissions from photon irradiation of samples stored in high extraction fields. However, only the low-energy electron emissions could be observed before energy filtering was recently introduced. Since then, energy filtering has made it possible to image certain photoelectron and Auger electron emissions using parallel mode imaging. A newly developed method called spectroscopic photoemission and low energy electron microscopy (SPELEEM)(Figure 7.12) integrates LEEM and XPEEM into a single synchrotron-based device(Aballe et al., 2019). By using an electron source, LEEM expands the range of characterization methods that can be used on the same sub-micrometer scale volume by adding low-energy electron diffraction (LEED), among other capabilities. This type of analysis is often called spectromicroscopy.

The reason for the higher spatial resolution is that, at high fields, electrons can be thought of as coming from point sources, such as atoms or ions that generate photoelectron radiation. This also applies to secondary electrons and Auger electrons, as both are produced by the relaxing of the core hole. When used properly, X-ray photoelectron spectroscopy (XPS) can yield a multitude of data regarding the elemental composition and speciation of a solid surface.

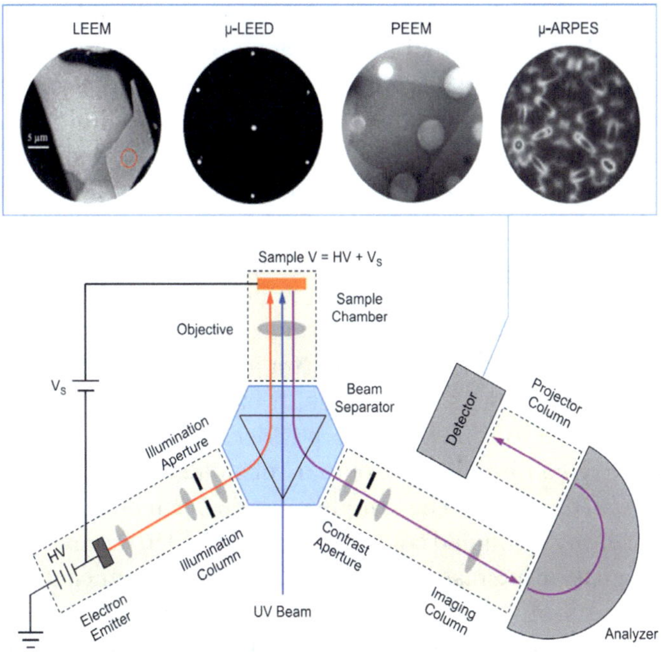

Figure 7.12. Spectroscopic photoemission and low energy electron microscopy (SPELEEM).

Source: Jin, W., & and Osgood, R. M. (2019). Spectroscopic photoemission and low-energy electron microscopy studies of the surface and electronic structure of two-dimensional materials. *Advances inPhysics: X, 4*(1), 1688187. https://doi.org/10.1080/23746149.2019.1688187, Creative Commons License.

7.12. SAMPLE HANDLING

Understanding the best sample handling practices to minimize contamination and/or changes to the sample surface of interest, the best data gathering techniques, as well as the best reference processes, is necessary to use XPS correctly. The extent and kind of damage that the sample may sustain during analysis, the volume examined and how it might be modified to meet the requirements of the particular experiment; and quantification techniques, including background subtraction procedures and possible error sources are crucial while dealing with XPS. XPS involves minimal sample preparation, in contrast to the majority of analytical procedures. The only requirements are that the sample be the right size for the instrument, the surface of interest be clean both before and during analysis, and the sample surface be compatible with X-rays and ultrahigh vacuum (UHV). Samples should ideally have the dimensions and shape of a small coin in terms of size. This is recommended because it is manageable and has an enough surface area to permit pump down speeds. In order to maintain their costs within reasonable limits, commercially available instruments also have access ports of a restricted size. Clips can be used to attach samples that are a few millimeters or bigger to the sample stage. Some other kind of support is required if they are powders or smaller than a few millimeters. Using indium foil strips, into which the sample is squeezed with a spatula or another piece of indium foil, is one method. After that, the strips are clipped onto the platform or sample stage. Another method is to apply the powder to double-sided stick tape of spectra grade that has been laid out on a flat disk and attached to the sample stage. It is always best to use pressurized N_2 or invert the stage to remove loose powder. In order to investigate nano-size powders, they can be pipetted onto a substrate, allowed to dry, and then dispersed in an inert solvent. The sample must be clean both before and during analysis because, any unwanted contamination applied to the area of interest will suppress the signal of interest. Gloves and cleaned instruments can guarantee cleanliness during sample preparation. Polyethylene gloves are recommended even if latex gloves can be used since some latex gloves contain silicones that, if injected, can segregate to the surface of a sample. Acetone, methanol, or isopropyl alcohol are typically used for cleaning. Similar to the tools required to make sure the sample is the right size to fit into the instrument, the instruments used to mount these samples onto the appropriate sample platforms must also be maintained clean. Samples should never be handled with the bare hands.

The samples are placed in an introduction chamber after they have been mounted. These are small enough to enable rapid pump down; for clean samples with low porosity, pump down periods is usually between 10 and 15 minutes. It is best to pump porous samples down in the introduction chamber for extended periods of time. After a pressure in the 10^{-7} Torr range is achieved, samples are moved to the analysis chamber. introduction chambers enable the quick transfer of samples from the atmosphere to the analysis chamber while preserving the UHV conditions necessary for analysis. UHV regulates the readsorption of gas-phase impurities onto the surface of interest during analysis, which is one of the reasons it is necessary in XPS(Senanayake et al., 2007). Only when cooling is used can high vapor pressure samples be evaluated under the necessary UHV conditions. Usually, an externally located LN_2dewar that is thermally coupled to the sample stubwhile inside the instrument is used for cooling. Liquid He must be used for further cooling. In certain conditions, the sample can also be heated while in the analysis chamber, or it can be rinsed in particular solvents or reagents before being placed in the introduction chamber to remove unwanted surface layers. Since carbonaceous overlayers easily generate CO_2, heating in the presence of a low-pressure O_2 atmosphere can also be helpful in eliminating them from thermally stable materials. As an alternative, sputtering can be used to remove particular surface layers in the analysis chamber right before analysis. There are a number of choices, some of which cause less damage to the sample than others. Sputtering with low-energy Ar^+ ions (500 eV), can eliminate surface impurities with minimal damage. Another approach that has drawn more interest is sputtering with cluster ion beams because it causes less damage when removing organic surface layers from organic substrates. It is recommended to gather spectra throughout a broad energy range, i.e., 0–1000 eV or more, when evaluating a sample of unknown origin and/or composition.

To maximize signal intensities while minimizing analysis times, this should be done with a step density of ~0.5 eV or larger under relatively low energy resolution conditions. The best emissions for use in quantification and/or high-resolution investigations are revealed by the spectra, which are frequently called survey spectra. High-resolution spectra needed for speciation investigations are obtained with the step density set at 0.2 eV or below and under high energy resolution conditions. Narrower energy ranges are chosen to compensate for the longer analysis times needed. Where feasible, interference-free emissions should be examined in order to reduce problems. Because switching the source from Al-Kα to Mg-Kα causes all Auger peaks to move along the $B.E._{XPS}$ scale by an amount equal to the energy difference of the source used, it is possible to eliminate any confusion regarding whether a signal originates from photoelectron or Auger electron emission. XPS uses the energy of the observed electrons to determine composition and speciation. This can only be accomplished using an energy reference. E_{vac} is the sole point of reference for electrons in gas phase atoms.

Damage to the material being examined is frequently observed during examination. This results from the fact that X-ray irradiation can cause surface-bound species to desorb, atoms or ions to be reduced inside the area under analysis, phase alterations within the sample under study, and migration of particular elements in specific samples caused by an electric field. When examining nonconductive materials and/or materials with loosely bound adsorbates, this is more visible. Exposing such materials to X-ray radiation will cause the charge in the area being examined to be removed and the electronic structure of the sample being studied to be disturbed. If the photoelectron producing atom or ion can exist in multiple oxidation states, then damage resulting from the elimination of charge can be understood. Spicer's three-step model illustrates how the intensities of photoelectrons rely on a variety of factors that can be separated into three fundamental groups(Suga et al., 2021). These group-specific characteristics consist of those that rely on the physics of photoelectron production, those that explain how the generated photoelectrons interact with the surrounding solid; and those that rely on the geometry of the instrument.

7.13. PHOTOELECTRON

The photoelectron cross section or probability often include the factors that are reliant on the physics of photoelectron production. These cross sections explain the probability of producing photoelectron energy from a particular core level of a particular element when exposed to photons of a particular energy. Thecrucial point here is that these are independent of the oxidation state and other chemical conditions of the photoelectron producing atom or ion. The characteristics that cause a decrease in photoelectron intensity are those that characterize how photoelectrons interact with the surrounding solid-state environment. The mean distance an electron with a given kinetic energy (K.E.) may travel in a given material before interacting and losing energy to its surroundings is described by these characteristics, which are typically represented by the photoelectron inelastic mean free path (IMFP).

The signal will blend into the background if this energy loss exceeds the energy resolution that was employed. This results from the fact that energy loss is usually non-discrete in nature. The likelihood that a photoelectron will be detected when it leaves the solid surface, how the relative photoelectron intensities depend on the X-ray source and the electron emission collecting angle; and the relationship between photoelectron intensities and, if relevant, takeoff angle and crystal orientation are considered in the setup. For a specific set of analytical conditions, this will stay constant even though it will depend on the conditions utilized. Variations in photoelectron intensities on the unpolarized photon (X-ray) energy and angle are described by the angular asymmetry factor. Since the X-ray source to electron collection angles are fixed, this is element and level specific; the instrument-specific dependence is constant, which is visible to the instrument operator because it is part of the quantification software. This relationship explains why different intensity ratios are observed when the same material is analyzed using devices with different geometries. Diffraction effects may occur when electrons move through a crystalline solid; this is the foundation of low-energy electron diffraction (LEED)(Figure 7.13). Along particular crystallographic directions, constructive interference will be observed if the Bragg diffraction conditions are met. Photoelectron intensities will increase along particular angles of emission.

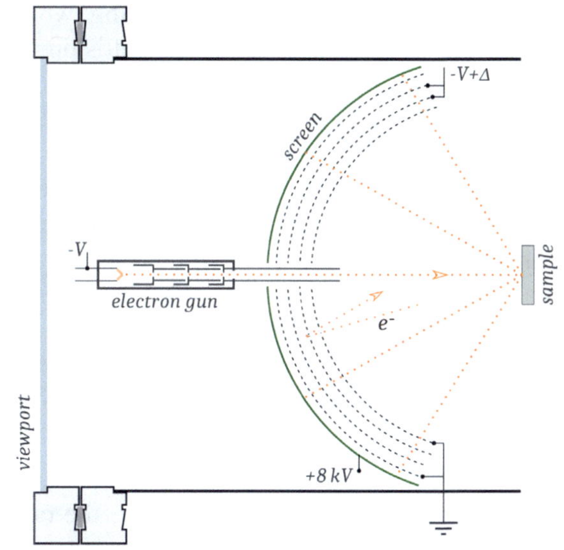

Figure 7.13. Low-energy electron diffraction (LEED).

Source: https://upload.wikimedia.org/wikipedia/commons/7/79/Low-energy_electron_diffraction_%28LEED%29_instrument_-_schematic_diagram.svg, Creative Commons License.

If not taken into consideration or filtered out, the intensity spikes added to the angle distribution will create errors in XPS quantification, even if they can be helpful for understanding structure. If analysis is done at an angle that is not typical, electrons that are experiencing elastic scattering can radiate from depths deeper than those suggested by their IMFP. The photoelectron cross section is the likelihood that a photoelectron will be released from an atom or ion after photon irradiation. However, given a specific set of analytical parameters, such as X-ray energy, photoelectron cross sections do not change. This is the cause of the relative simplicity of quantification in XPS when employing Al-Kα and Mg-Kα X-rays. These are sometimes referred to as Scofield cross sections because Scofield tabulated them for all elements from Li–U under Al-Kα and Mg-Kα irradiation(Nefedov et al., 2008). Before an exact quantification can be reached, the sensitivity factor associated with each source must be determined if other sources are to be used. The distance a photoelectron can travel within a solid before losing energy, or before experiencing inelastic collisions as defined by the IMFP, is one of the main parameters influencing the volume under analysis. Elastic and inelastic scattering, with the latter being the predominant effect, determine how far a photoelectron can travel in a solid. Collisions where no energy is exchanged are described by elastic scattering, while collisions where energy is exchanged are described by inelastic collisions. Mean or average values are employed since inelastic scattering is statistical in nature. The average distance traveled by an electron of a given energy within a single-layered homogeneous amorphous solid between two consecutive inelastic scattering events is known as the IMFP. The attenuation length (AL) is the average distance, including elastic scattering, that an electron of a given energy travels between two consecutive inelastic scattering events within a certain multilayered and/or heterogeneous material. The 'Information Depth" indicates how far below the surface a given proportion of photoelectrons originate. The depth from which 95.7% of all photoelectrons originate is known as the sampling depth. The distance normal to the surface from which 61.7% of the initial photoelectron population arises is known as the 'Escape Depth.'

When defining the concentrations of elements within a surface or subsurface layer, the information depth and sampling depth are the ones that must to be used. The statistical nature of the causing scattering events is best described by these and the escape depth, which show that the population of electrons with a given energy emitted from a solid decrease exponentially with their depth of origin. This makes it clear that there is a chance that electrons with a given energy could escape from a depth than would be predicted from their IMFP. Compared to photon-electron interaction, the likelihood of electron-electron interaction is higher. This explains why methods that analyze accompanying photon emissions, like X-ray fluorescence (XRF) and energy-dispersive X-ray analysis (EDX), are not surface specific, whereas methods that record electron emissions, like XPS and AES, are. The popular approaches for determining the probability of electron-electron interactions are empirical or semiempirical. These show that contact probabilities follow an exponential function for electrons with K.E. values greater than about 100 eV. The Beer–Lambert law, which was restated by Lambert and then reconstructed by Beer from the original discovery by Bouguer, can be used to approximate the path length of electrons with respect to the initial direction of movement. This yields an average value with a Poisson distribution. The distance over which a signal intensity decreases when represented on the natural log scale or by 61.7% when plotted on a linear scale is the path length, and hence the parameters AL and λIMFP(Flores-Mancera et al., 2020). An intensity reduction of 85.6% corresponds to e^{-2} and 95.7% to e^{-3}. A 61.7% loss corresponds to the escape depth and a 95.7% loss corresponds to the sampling depth when non-normal takeoff angles are used to define λIMFP. The distinction between AL and λIMFP is that the former takes into account both elastic and inelastic scattering effects, while the latter only takes into consideration inelastic scattering. Elastic scattering is included because, when layers of varying electron densities are analyzed at angles other than takeoff, it can increase the apparent depth from which electrons can appear to originate, even though it does not directly cause any change in energy.

7.14. BACKGROUND NOISE

Managing the background signal is one of the main challenges in measuring XPS spectra. The spectra obtained from transition metals is one example for this. There are several origins of the background signal, but inelastic electron scattering is common. A broad background to the lower $K.E._{XPS}$ side of the peak of interest is produced by the non-discrete energy loss via inelastic scattering, which is energy lost by electrons still inside the solid to its surroundings. The backdrop of peaks of interest can be described and eliminated using three popular subtraction techniques. These are the Shirley, Tougaard, and Linear routines. A linear function drawn between two points on either side of the photoelectron peak of interest is assumed by the linear procedure. Despite being an extremely basic method, it can yield results. The popular approach in the early days of XPS was this one, mostly because of computing constraints at the time. Peak intensities and the intensity trend observed on the lower and higher $B.E._{XPS}$ sides of the corresponding photoelectron peak define the form of the Shirley routine, which assumes an S-shaped background for each peak.

A repeatable, methodical methodology is necessary for quantification in order to achieve the necessary precision. Accuracy errors are then taken into consideration by applying the proper sensitivity factors. Only the Tougaard procedure makes an effort to compute the background, or to model the active inelastic scattering events, albeit within a constrained energy range. Because of this, this is the complicated of these three methods but also the least subjective. This background may be fitted over the relatively small energy range that is usually of interest due to the development of a universal loss function, which attempts to simplify the process. A less popular but incredibly precise method is to compute the background from reflected electron energy loss spectroscopy (REELS)(Figure 7.14)spectra that were taken with the same device. In order to study the region of interest, this AES-derived approach necessitates an energetic electron gun with sufficient energy. The capacity of XPS to quantify recorded signals with relative ease and without the need of reference materials is one of its advantages. In these conditions, one can anticipate an accuracy of no more than 5% to 10%, with errors of up to 30% occurring if the background is not adequately taken into consideration, the electron IMFP values are inaccurate, within the examined region, there is a prominent difference in concentration. This discrepancy could result from small inclusions or macroscopic areas with varying compositions.

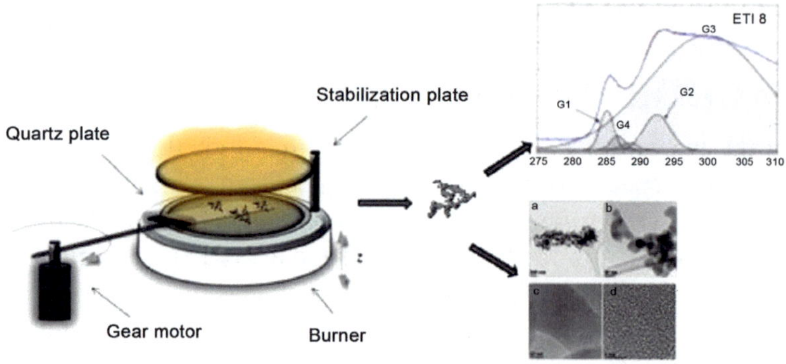

Figure 7.14. Electron energy loss spectroscopy.

Source: Russo, C., Pawlyta, M., Tomiczek, A. E., Ciajolo, A., &Apicella, B. (2021). On the application of electron energy-loss spectroscopy for investigating nanostructure of soot from different fuels. *Fuels, 2*(3), Article 3. https://doi.org/10.3390/fuels2030021, Creative Commons License.

CHAPTER 8

BIOMOLECULAR SPECTROSCOPY

8.1. INTRODUCTION

Two-dimensional infrared (2DIR) spectroscopy (Figure 8.1) is a rapidly evolving technology. Because it spreads the transitions over a 2nd (second) frequency domain, it can provide more structural detail (resolution) than classical FTIR.

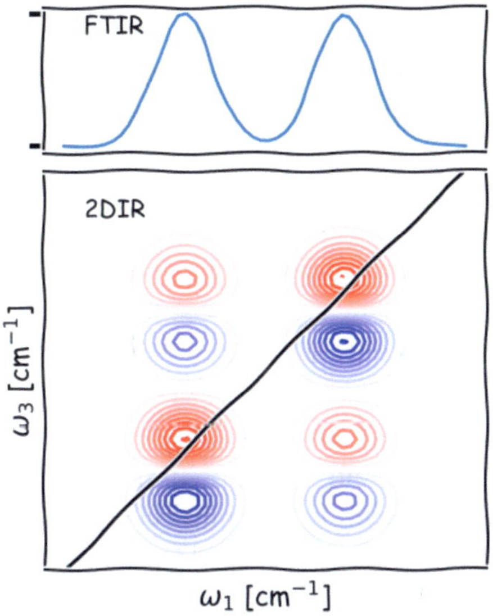

Figure 8.1. Two-dimensional infrared (2DIR) spectroscopy.

Source: https://upload.wikimedia.org/wikipedia/en/6/67/Schematic_of_typical_two-dimensional_infrared_spectrum.jpg, Creative Commons License.

This reveals cross peaks, which measure the coupling between molecular vibrations and other vibrational correlations. They provide useful information about molecule structure, dynamics, and solvent effects, which may be of significance. Cross peaks have the ability to clarify (resolve) information about 3D structure and aid in deciphering data in packed bands. A sequence of infrared (IR) pulses is used in 2DIR, a type of nonlinear optical spectroscopy, to interact with the vibrational transitions of the sample molecule or molecules. Similar to two-dimensional NMR techniques, this involves interactions between the nuclear spin changes and many radiofrequency signals. Because of the intrinsic timescales, 2DNMR typically provides a time-averaged image of the molecule in addition to comprehensive spectroscopic information. Although 2DNMR techniques can indirectly access timescales down to the picosecond level, this can cause sophisticated models to be generalized. Although it does so at the expense of the resolution offered by 2DNMR techniques, 2DIR can access periods ranging from picoseconds to seconds (Rubtsov, 2009).

2DIR provides the capacity to examine the dynamics in greater depth, even though 2DNMR is a strong tool for examining the structure of many systems. In order to get 2DIR data experimentally, two techniques have been developed: the frequency domain double resonance 2DIR approach and the time domain Fourier transform method based on vibrational photon echoes. By using a four-wave mixing methodology, both methods explore the same third order response function; three infrared pulses interact with the sample to create the intended nonlinear signal field. The Fourier transform relationship relates the identical data derived from frequency and time domain approaches. There are technological differences between the two approaches. A common pump-probe experiment serves as the foundation for the frequency domain approach. The laser is divided into four beams, which are positioned at three corners of a square and directed towards the sample in the photon echo method. The signal or echo pulse is sent in the direction of the square 4th.

In a 2DIR experiment, the sample is impacted by the three laser pulses at three different times. A coherent superposition or mixed state of the ground and first excited states is created by the interaction between the first pulse and the sample. The coherence evolution period (t_1) is the time after the first pulse; stimulated molecules oscillate in phase at first, but the initial phase coherence is lost due to minute changes in the frequency of molecular vibrations. This is called dephasing. The superposition is changed into a population state, either the ground state or the first stimulated state, by the second pulse. The waiting or mixing period (t_2) is the time after the second pulse. The length of t_2 can be changed during this time, and vibrational dynamics are seen. The composition of the superposition state, which is replicated in the third pulse, is determined by the type of population state that was occupied during t_2. The sample does not instantly regain coherence; rather, it goes through a rephasing process, during which the polarization linked to this leads to the emission of the echo pulse and coherence is recovered after a time comparable to t_2. The pump and the probe are the two pulses that happen during the mixing period. The probe is weaker than the pump beam, which is stronger; the two beams focus and overlap at the sample, and the time delay between them is adjusted.

8.2. PHOTON ECHO TECHNIQUE

The primary advantage of the photon echo technique over double resonance is that, because of its smaller pump pulse bandwidth, it preserves the time resolution that the latter loses. This results from the capacity to distinguish between the time orderings of the laser pulses and thoroughly examine the interface between the laser and the molecule. The photon echo spectrum is devoid of the distortions that arise in the pump pulse experiment due to the finite bandwidth, or the finite range of frequencies in the wave, even if the spectral peaks appear the same for both approaches. The amide I band, one of the primary vibrational modes in proteins, has a vibrational lifetime of less than two ps. Ground state bleach, stimulated emission, and excited state absorption are the different types of contributions that make up a 2DIR spectrum(Baiz et al., 2009).

When a single excited state and the ground state are coherently superposed by the initial laser pulse, the ground state bleach signal is produced. The system is brought back to the ground state by the second pulse. A third pulse produces the superposition following the waiting period (t_2). The system relaxes to the ground state from the superposition by emitting the signal after the third pulse. The system either resides in a single excited state or a superposition of two single excited states throughout the waiting period, but the stimulated emission is produced by the same series of pulses. The final laser pulse brings the system into coherence between double and single excited states, which is why the excited state absorption signal is different. Due to the involvement of doubly excited states, this contribution is susceptible to anharmonicities, or departures from harmonic vibrational motion; in the absence of these, the entire signal is lost. The locations of the overtones in an infrared spectrum are likewise caused by anharmonicities. The fundamental frequency and its integral multiples cause a vibration to oscillate. The oscillations above the fundamental frequency produce the overtones, and their position is altered by anharmonicity. The first overtone in an anharmonic system will be lower than the fundamental frequency, but in a simple harmonic system, it will be double that.

8.3. COMPUTATIONS

Theoretical computations help understand experimental spectra; this is particularly true for vibrational spectroscopy in biological systems. Because of the strong theoretical foundation, theoretical models like those derived from molecular dynamics (MD) simulations, can be used to compare experimental results. Both theoretical models and experimental techniques can benefit from these comparisons. Local structure and global conformation are the two primary experimental sources of information in a protein 2DIR spectrum. Site-specific isotope tagging can be used to identify the local structure composition, which includes solvent exposure and hydrogen bonding. Delocalized excitonic bands reveal details about the secondary structure as well as the global structure. A delocalized excited state known as an exciton is produced when many separate sites interact with one another. Any theoretical model for protein 2DIR should consider both global and local effects, including the delocalization of amide excitons brought on by site-to-site coupling and the impact of the local electrostatic environment on vibrational frequency. The sensitivity of oscillator site energies, or the frequency of each individual oscillator, to their local surroundings should also be described by a theoretical approach for 2DIR. In theory, ab initio electronic structure computations might be used to determine the exciton Hamiltonian for every time step in a MD simulation. High-level computations on the several configurations needed to produce the spectral line shape are not feasible. Frequency maps and ab initio-based coupling have been created to address this problem. These frequency maps clearly show how a particular chromophore frequency and the electrostatic environment are related. The Cho frequency map is an example. Many ab initio computations on solvated N-methylacetamide resulted in the creation of this map(Yu et al., 2010). It connects the peptide amide I vibrational mode frequency to its structure and solvation. The amide I modes of various polypeptides in solution, including 2DIR spectra, have been extensively calculated using the Cho frequency map. Protein (thermodynamic) stability, structure, and function are significantly influenced by flexibility and dynamics.

8.4. PROTEIN DYNAMICS

Accurately characterizing protein dynamics on extremely short timescales is one of the key challenges in biophysics. Because protein behavior happens over a wide range of timescales, from the picosecond to the second, no direct measurement that pinpoints the precise link between dynamics, structure, and function has been made. Conducting a single experiment that looks at all of these timescales at once is challenging. A lot of effort has been put into developing an image of protein dynamics on a picosecond timescale. Because 2DIR can resolve time in picoseconds, it might be possible to determine the precise nature of this relationship. When it comes to IR spectroscopy of proteins and polypeptides, the amide I, II, and III bands are the three main bands of interest. The amide I-III bands are commonly utilized, while there are others, such as the amide A band. Protein secondary structure and the amide I band are highly connected; different spectral signatures are produced by α -helical and β -sheet structures. The amide I mode is mostly a CO stretching mode, with a minor CCN deformation component and an out-of-phase CN stretch mode contributing slightly. It occurs in the spectrum 1,600–1,700 cm^{-1} range and, in the case of FTIR, is typically wider and crowded or congested, and challenging to understand because of the high number of vibrations in this region. The primary components of the amide II band (1,510–1,580 cm^{-1}) are a CN stretch and an in-plane NH bend. Smaller contributions from a CO in-plane bend, a CC stretch, and an NC stretch make up the out-of-phase combination that contributes to the amide II band. Since the connection between the amide II band and secondary structure is less clear, it is not as frequently utilized as the amide I band, while being sensitive to protein structure. This band has been employed in 2DIR investigationsand is mainly recognized for its sensitivity to the protonation state of peptide units. NH in-plane bends and CN stretches combine to form the amide III band (1,200–1,350 cm^{-1})(Azeez and Shenbagaraman, 2025). A CO in-plane bend and a CC stretch also contribute to it. The amide I band is more helpful for secondary structure investigation than this mode, which only results in very weak bands in IR spectra. The amide III band is too low in frequency to be efficiently probed using ultrafast laser equipment, and it is located in a complicated spectral region. Secondary structure, individual peptides or small proteins, amyloid fibrils, and membrane proteins are thebroad categories into which recent applications of 2DIR to proteins belong. Since the amide I band contains information, 2DIR is ideally suited to study the structure of proteins and peptides. Because of the static distribution of various frequencies resulting from these local fluctuations, these vibrations are susceptible to local differences in structure, which causes an inhomogeneous widening of the line shape.

8.5. VIBRATIONS

Researchers still do not fully understand the precise nature of the interaction between the structures causing these vibrations and the frequency and amplitude of IR bands. But by distributing the spectra across a second frequency axis, 2DIR removes some of this congestion and makes it possible to gather more structural and dynamic data. Gaining a deeper comprehension or understanding of how secondary structural components result in detected spectrum indications would have significant ramifications for ligand binding, drug development, and folding pathway understanding. NMR, X-ray crystallography, and cryo-electron microscopy are methods for examining the structure of proteins. X-ray crystallography has made it possible to determine the structures of many proteins. This approach cannot solve the structure of many proteins because they are difficult to crystallize. NMR is sensitive to the structure, heterogeneity, or conformational dynamics of proteins and is frequently employed to solve protein structures. Several protein groups, such as fibrous proteins, naturally disordered proteins, gels, amyloids, and aggregates, have proven challenging to define despite the abundance of methods available.

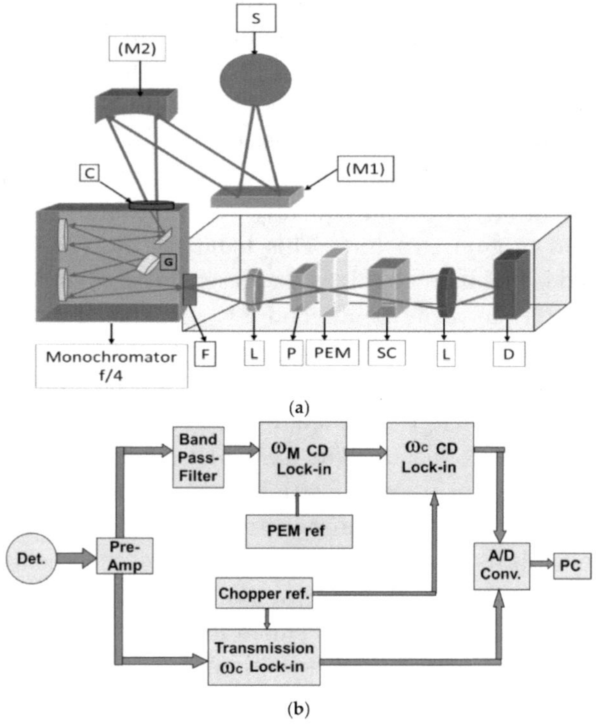

Figure 8.2. Electronic circular dichroism (a and b).

Source: https://upload.wikimedia.org/wikipedia/en/6/67/Schematic_of_typical_two-dimensional_infrared_spectrum.jpg, Creative Commons License.

For these proteins, a technique that measures the proportion of amino acids in α-helices or β-sheets would be especially helpful. Based on the distinct fingerprints of secondary structure elements, a novel technique for measuring the structure of proteins in solution has been developed utilizing

2DIR. According to the approach, each distinct secondary structure has its own spectrum, and the spectrum for the entire protein is the sum of these contributions. These presumptions are not totally accurate given what we now know about vibrational coupling and secondary structure spectrum overlap. Because 2DIR offers more distinct spectrum information than FTIR, it is a more reliable predictor of structure. The model ability to quantitatively ascertain the secondary structure content of proteins without the need for isotope labeling was demonstrated by testing it on commercially available globular proteins with well-defined structures, such as myoglobin, insulin, and ubiquitin.

The errors related to electronic circular dichroism (CD), (Figure 8.2)a common technique for determining protein conformation in solution, and 2DIR structure determination were similar. Although CD experiments are simpler to conduct than 2DIR, the latter may provide valuable information. 2DIR could be used in conjunction with a laser-induced temperature jump to track protein folding/unfolding in solution by taking use of its quick time resolution and the capacity to isotope label specific residues to improve structural resolution. The applicability of isotope-labeled 2DIR for acquiring site-specific structural data on Trpzip2 has been examined theoretically(Smith et al., 2010).

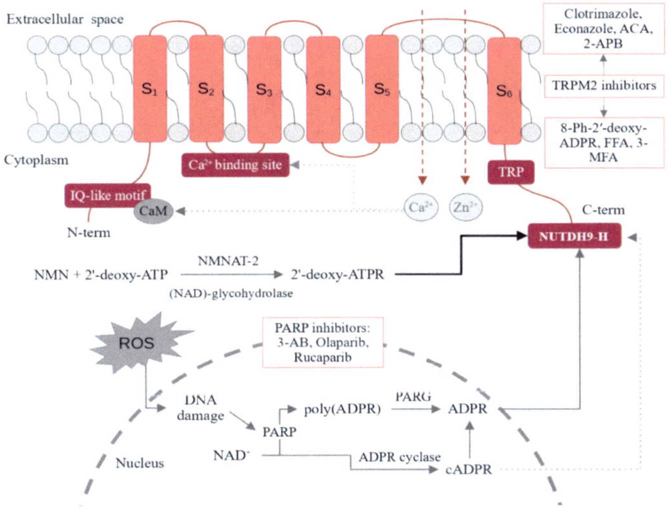

Figure 8.3. Trpzip2.

Source: Zielińska, W., Zabrzyński, J., Gagat, M., &Grzanka, A. (2021). The role of TRPM2 in endothelial function and dysfunction. *International Journal of Molecular Sciences, 22*(14), Article 14. https://doi.org/10.3390/ijms22147635, Creative Commons License.

Trpzip2 (Figure 8.3)is a synthetic β-hairpin peptide that was created especially as a model system for research on β-sheet protein folding(Figure 8.4). It has the amino acid sequence SWTWENGKWTWK and four fluorescent tryptophans. The hybrid zipper model, zip-out, hydrophobic collapse, and reptation are the four folding mechanisms that have been proposed. According to the zip-out model, the turn forms first, then the interstrand hydrogen bonds gradually form toward both termini, and finally the hydrophobic core forms at the end. According to the hydrophobic collapse model, interstrand hydrogen bonds develop in both directions once the hydrophobic core forms first.

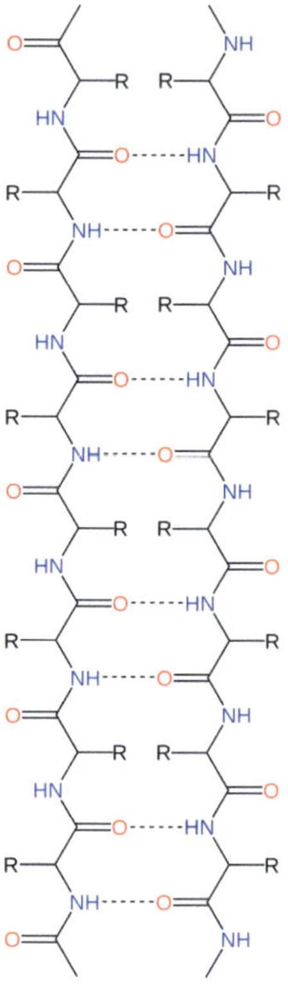

Figure 8.4. Beta-sheet protein folding.

Source: https://en.wikipedia.org/wiki/Beta_sheet#/media/File:Beta_sheet_bonding_antiparallel-color.svg,Creative Commons License.

Non-native hydrogen bonds are first established with the native state toward the end of the slide in the reptation mechanism, which involves the β-sheets sliding. According to the hybrid zipper model, the turn forms first, then non-polar residues collapse into a hydrophobic core, which is stabilized by terminal salt bridges. Last to form are the hydrogen bonds between the strands.

Trpzip1 and Trpzip3 are related zippers that have distinct folding rates despite simply having a different turn structure. This implies that a crucial stage in the folding process is the creation of the turn. Various peptide residues have been the subject of many isotope labeling investigations. Because of its higher mass, the carbonyl vibrational frequency is red shifted, creating discrete peaks in the spectra. However, to distinguish between the suggested folding mechanisms, temperature-jump tests on only a few labeled sites would be insufficient. Some researchers concentrated on every amide I site in the peptide. MD was used to simulate the proteins, and the vibrational Hamiltonian form was developed each snapshot (10 fs) in the trajectory. Instead of an electrostatic map, a nearest-neighbor frequency shift map was used to correct the site frequencies, which were discovered using a Stark effect-based technique. The NISE approach was used to compute the 2DIR spectra from the time-dependent Hamiltonian(Jansen and Knoester, 2009).

To determine their usability in categorizing the amide I groups of Trpzip2, the fluctuations of a number of spectral line shape parameters were examined, including the diagonal and antidiagonal line width of the 2DIR peaks. The amide I groups were identified by the standard deviation of their amide I frequency variations and were categorized as terminal, internal, external, and turn based on where they were in the hairpin. The diagonal and antidiagonal peak widths at a single location within the 2DIR spectrum are connected to this fluctuation. Sites exposed to water had wider antidiagonal widths, whereas sites with carbonyl groups buried in the middle of the protein had narrower antidiagonal widths. The electric field produced by the water acting on the amide group is responsible for the line width. A technique for distinguishing between internal and solvent-exposed locations is provided by this measurement of the line forms. An isotope-labeled 2DIR spectrum antidiagonal line width of peaks could be helpful in examining the folding kinetics of Trpzip2 and other peptides. The spectral line will broaden (widen) as a result of carbonyl groups that were previously inside the peptide being exposed to water during unfolding. The suggested zip-out, hydrophobic collapse, hybrid zipper, and reptation folding mechanisms should all be distinguishable by labeling the interior carbonyl sites and carrying out temperature-jump studies.

8.6. SHEETS

Experimental tests were conducted on theoretical models for coupling between residue pairs in parallel β-sheets, and the impact of the environment on coupling was examined. To address the problems with employing natural β-sheets, the relative scarcity of parallel β-sheets and that small peptides do not readily form the structure, a synthetic peptide macrocycle was employed. Each of the two strand-forming segments that made up the macrocycle had eight L-α amino acids. The strands are joined by a D-prolyl-1,2,-diamino-1,1-dimethylethyl unit at the C-termini and a glycyl-succinyl unit at the N-termini. These terminal linkers encourage β-sheet interactions between strands. Since vibrational spectroscopy is crucial for examining the structures of amyloid fibrils, (Figure 8.5)which consist of β-sheets, it is crucial to have a detailed understanding of parallel β-sheet IR spectra.

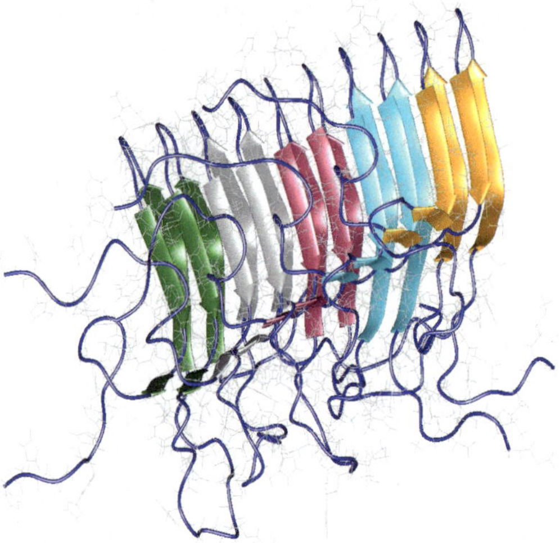

Figure 8.5. Amyloid fibrils.

Source: https://upload.wikimedia.org/wikipedia/commons/4/45/2rnm.jpg, Creative Commons License.

Although there are already a lot of experimental and simulated spectra for antiparallel β-sheets, Woy study on parallel beta sheets is crucial. The findings evaluate the coupling theories that have been applied to the interpretation of protein β-sheet IR spectra for the past 50 years. Variations caused by external influences are on par with or greater than the intrinsic connections (bonds) found in secondary structure. The secondary structure frequency distribution is not significantly wider than random coils. Since even a simple coupling alters the intensities, the main effect of coupling is the redistribution of oscillator strength. Prior research on nitrile groups showed that while small couplings had an impact on intensities, they had no effect on frequencies. More precise intensity calculations should be possible if the transition dipole orientation with regard to oxygen, carbon, and nitrogen is improved. This study helps make FTIR and 2DIR more quantitative instruments for structure elucidation and increases confidence in the correctness of calculations of solvent-dependent spectral characteristics, such as line widths and frequency shifts.

8.7. PROTEIN FOLDING

In aqueous solution, even small polypeptides only take on a small number of conformations. Protein folding pathways may become stronger if the conformations of these small peptides are understood. The development of local secondary structure, which may act as a nucleation site for the folding of the entire protein, may be one of the initial stages of a folding pathway. If the conformational preferences of short peptides and their reliance on environmental parameters, such pH or temperature, could be systematically studied, the location of such nucleation sites might be predictable. A strong tool for carrying out this investigation is provided by 2DIR. Using 2DIR and an excitonic model for the coupling of the amide I modes, the structures of the tripeptides L-alanyl-L-lysyl-L-alanine and L-alanyl-L-histidyl-L-alanine have been investigated(Huerta-Viga et al., 2010). Measurements in D_2O are indicated by the prime. The dihedral angles (φ,ψ) of the backbone are not substantially impacted by the protonation of side chains or C-terminal groups. This suggests that the creation of secondary structures for these peptides may not be hindered by low pH. The data was analyzed using the angle between the transition dipole vectors and the couplings between the amide I modes. The upper and lower bounds for the dihedral angles were determined using the confidence limits of these variables. Confidence regions for the tripeptide conformations in the map were calculated from these limitations. A generalization of confidence intervals applied to multiple variables at the same time is called a confidence region. A complex confidence region resulted from the extremely nonlinear interaction between the dihedral angles and the observed couplings and transition dipoles. This could affect subsequent research because similar considerations will apply to any technique that uses intramolecular coupling. On a de novo 17-residue α -helix, MD simulations and theoretical 2DIR calculations have been performed. It is a newly created peptide that forms an α -helix on its own without the assistance of disulfide bonds. In order to record the folding event, researchers began with an extended structure of the peptide and followed a single trajectory in a solvent and at room temperature. A standard, nonpolarizable force field and a force field that substituted the adaptive hydrogen bond-specific charges (HBC) scheme for the atomic charges were the two force fields used to create the trajectory. Residues not involved in hydrogen bond production or breaking were excluded from the HBC design, which addressed these processes by incorporating their polarization into the folding process. Using the polarizable force field and the HBC method, folding happened in 10 ns. Although partially folded objects were seen, insufficient hydrogen bonding prevented a folding event in the nonpolarizable force field. The MD simulations were used to compute the 2DIR spectra, which were then averaged across 1,000 trajectory snapshots. The frequency of the vibrational excited state, the homogeneous dephasing rate, and the vibrational transition dipole from the ground state to the excited state are used to compute the linear absorption.

8.8. MAP

An electrostatic map was used to determine the frequency of the vibrational excited state and the vibrational transition dipoles(Gruenbaum et al., 2013). The nonlinear exciton equations were solved to produce the optical response utilized in the 2DIR computation. This peptide simulated folding complied with the funnel free energy landscape theory of protein folding. The mechanism used downhill folding kinetics to fold itself. Information that was previously unavailable through other experimental techniques was made available by the 2DIR simulations. Both theoretical and experimental 2DIR research have previously concentrated on native or denatured protein structures. The 2DIR simulations yielded residue-specific information when used in conjunction with isotope labeling. Changes in the line shape and location of the cross peaks in the 2DIR spectra are indicative of the development of backbone hydrogen bonds and the alpha-helix that follows. Although the experimental 2DIR spectra are still pending, the folding process should be empirically followed as 2DIR develops. The majority of recent 2DIR research on protein folding has concentrated on photo-switchable proteins or the unfolding process utilizing a temperature-jump technique. The MD simulation structural and dynamical data can be used to draw conclusions, but the accuracy of the simulations can be confirmed by computing the 2DIR spectra. A peptide with 35 amino acid residues is known as chicken Villinheadpiece 35 (HP35). Several computational and experimental investigations have used it as a model. It is a subdomain of the F-actin binding protein. Despite its small size, this protein possesses a number of traits often found in larger proteins, including significant resistance to different point mutations, autonomous folding, and a hydrophobic core that stabilizes the tertiary structure. It has been a choice for characterization using

MD simulations and experimental methods like fluorescence resonance energy transfer (FRET) because it is one of the fastest folding peptides. Nitrile (CN) groups bonded to the peptide have recently been employed as a vibrational dynamics label in 2DIR analysis of HP35. One (HP35-P) and two (HP35-P_2) cyanophenylalanines in the hydrophobic core of HP35 variants were examined. The kinetics and absorption spectra of the two variations as determined by 2DIR were comparable within experimental error. Adding the CN label by replacing phenylalanine with cyanophenylalanine slightly alters the system dynamics and structure, and this should also be the possible for larger proteins.

The dynamics of HP35 and its fast-folding mutant HP35 NleNle were measured using the nitrile label. Due to the substitution of nonpolar amino acids for two charged amino acids in its hydrophobic core, HP35-P NleNle folds slowly. According to research on unlabeled peptides, the mutation produced a more stable peptide as determined by temperature-dependent CD for the CN-labeled peptides. Therefore, the altered structure of the folded peptide may be the cause of the mutant improved stability and faster folding time. For additional MD simulations, HP35-P and HP35-P NleNle would both be good options. Due to a lack of appropriate structural tools, proteins that are intrinsically disordered cannot be crystallized, and X-ray diffraction and scattering experiments provide insufficient information. Although it can only examine long-lived conformational states, NMR is a promising technology that is unable to completely resolve the heterogeneity of these systems. Studying disordered systems and identifying unique conformers and structural variants are ideal uses for 2DIR sub-picosecond temporal resolution. Transport proteins, a class of membrane protein, have many subcategories and are responsible for moving ions and other small molecules across the membrane. Many biological functions, such as photosynthesis, enzyme catalysis, and acid-base neutralization, depend on proton transport.

Hydronium ion production is one way that proton conduction might take place. This makes it possible for an excess proton to pass through the cell membrane by using the hydrogen bonding network of water molecules inside the channel. The constricted circumstances inside the protein hydrophilic cavity allow for the transfer of the proton through the production of a hydronium ion, which changes the structure of the water hydrogen bond network to resemble a wire rather than the structure found in water. The foundation of 2DIR investigation of proton transport is that changes in the amide I vibrational frequency are caused by the dynamics of the protein channel and are also influenced by proton transfer. Proton transport channels, the line form of the amide I band, and the degree to which 2DIR can probe the proton transport itself were all examined by researchers. They investigated the proton channel, the peptide gramicidin A.

The enhanced permeability of inorganic monovalent cations across bacterial cell membranes is the source of Gramicidin A antibacterial action. The proton/ion gradients between the extracellular and intracellular environments are destroyed as a result. Proton hopping or water reorientation are the two possible ways via which protons are transported in the channel. To investigate proton transport via the channel, MD simulations and 2DIR calculations were employed. The amide I site frequencies were determined using the Stark effect approach in conjunction with the OLPS-AA force field.

8.9. NISE AND 2DIR

The NISE methodology was used to calculate the 2DIR spectra from the Hamiltonian. By tracking the time evolution of the antidiagonal line width and spectrum slope, the experiment demonstrated that proton transport does alter the system dynamics. The factors at play, such as channel water rotation and proton hopping, cause the position-dependent proton transport rates to be smeared out in the measured isotope tagged 2DIR spectrum. Within roughly 10 A^o, the local amide I site frequencies are sensitive to proton hopping. According to the simulations, a proton entering the channel seems to be the rate-limiting step. The proton has a limited chance of entering the channel from the bulk region, but it can readily escape back to the bulk from the channel. Therefore, a proton will only stay inside the channel for a considerable length of time at low pH, and 2DIR will only be sensitive at low pH when exploring proton transport in membrane proteins. Studies are also being conducted on other proton transport-related systems, including fuel cell electrolyte membranes, condensed phase crystals, carbon nanotubes, and enzyme families. A potential method for researching proton transport in these systems is the combination of 2DIR spectrum modeling with MD simulations. The intermolecular proton transport rate in biological channels may be experimentally determined by measuring the spectrum line width as a function of waiting time. Researchers discovered that the proton hopping rate and the antidiagonal line width temporal evolution rate are not equivalent.

8.10. PROTON TRANSPORT

More sophisticated proton transport model-based techniques should be able to confirm the association their model demonstrated between spectral behavior and the actual proton transport rate(Kornyshev et al., 2003). Although their approach was parameterized for residue four of gramicidin A, it is anticipated that the proton and water electric fields will account for the majority of frequency fluctuations, with contributions from the particular site being minimal. Other proteins involved in proton transport should be able to use the technique. Researchers examined membrane proteins that are involved in the transfer of ions rather than protons. On specific temporal and energy scales, protein complexes may exhibit non-classical phenomena like coherent excitation transmission. There are still concerns about whether the reported behaviors happen in other systems, despite the fact that a lot of theoretical and experimental work has been done, particularly on the photosynthetic system. Quantum impacts could provide fresh insight into the underlying ideas guiding these system operations. One of the well-researched ion channel classes is the potassium ion channel. Many explanations for the high transport rate and selectivity have been proposed using crystallography and molecular modeling. The snug-fit model of ion selectivity is the most popular. The channel protein was subjected to MD simulations, and the trajectories that emerged were utilized to compute the 2DIR spectra. This protein was the largest for which simulated 2DIR spectra had been computed at the time. Due to rigidity from ion coordination, preliminary research on non-actin, valinomycin, and potassium revealed spectral line narrowing. When potassium is present in the channel, the ion channel itself exhibits the same line narrowing. It is possible that the intensity variations are enough to distinguish potassium binding. The frequencies of oscillators coordinating potassium changed in tandem with the stiffness brought on by potassium binding. It is challenging to identify this alteration in the native channel protein. These modifications can be emphasized by isotope labeling, which also generates differential signals that can be used to detect switching. The main reason for the vibrational spectrum fluctuations that were detected was electrostatic alterations or changes. The electrostatic alterations brought on by the potassium ion mobility account for a larger portion of the observed spectrum variations than do local modifications such carbonyl group rotations. The in-plane ring stretching vibrations over the pyrimidine and purine rings are significantly delocalized, as demonstrated by 2DIR. For certain vibrations, like the carbonyl stretches on guanine and cytosine, such assignments are still reasonable.

8.11. STRETCHING

Studying the N–H stretching modes using vibrational spectroscopy offers a direct probe for the base pair interactions since the complementary nucleic acid base pairs interact through hydrogen bonding. The diagonal and off-diagonal anharmonicities are not stated, and the assignment of these bands using conventional linear spectroscopies is still unclear. Studies on the N–H and C = O stretches of DNA show that 2DIR provides direct insight into such behavior. In spite of this, diagonal and off-diagonal anharmonicities have yet to be discovered. Researchers combined computed 2DIR spectra of DNA with MD simulations. Two double helices, dG_n–dC_n and dA_n–dT_n in a B-DNA structure with a helical twist of 36° and an interlayer spacing of 3.4 A°, were specifically taken into consideration for the simulations. Solvation dynamics and conformational fluctuations were examined from these simulations. For the oligomers under study, the structural differences in the bases caused comparatively less variance in the frequencies and coupling strengths in water. When there were more than five base pairs in the strand, the spectra for both dG_n–dC_n and dA_n–dT_nslightly varied with the number of base pairs(Dadarlat and and Saxena, 1999). The 2DIR spectra of the A-, B-, and Z-forms of DNA were simulated by Lee and Cho using quantum chemistry computations. The DNA conformation had a significant impact on the vibrational coupling constants. The Z-form of DNA exhibited the higher interlayer distance among the three types. The majority of the vibrational coupling constants between adjacent base pair layers are therefore lower in Z-DNA (Figure 8.6)than in the other forms. The Z-form has highly concentrated vibrational modes, whereas the A-DNA exhibits rather substantial vibrational delocalization. This can be investigated by looking at how the FTIR and 2DIR spectra vary with chain length. Compared to the spectra of A-DNA, the Z-form of DNA vibrational spectra are substantially less reliant on chain length.

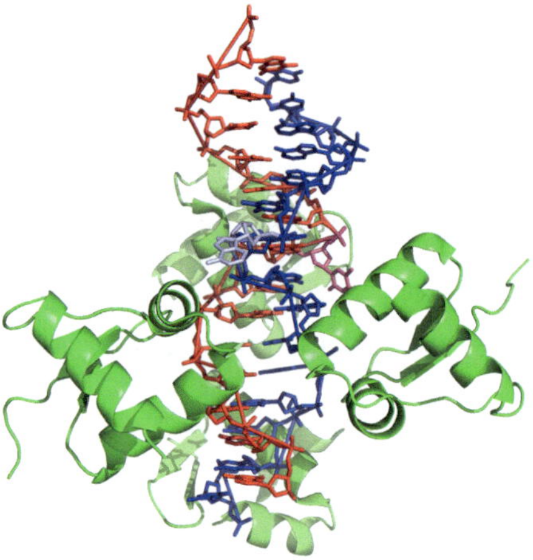

Figure 8.6. Z-DNA.

Source: https://upload.wikimedia.org/wikipedia/commons/5/57/B-%2C_Z-DNA_junction_2ACJ.png, Creative Commons License.

To test a novel coupling model for the carbonyl stretches of guanine and cytosine, researchers used two model DNA oligomers. The two oligomers that were employed were $d(GC)_8$ and double-stranded $d(G_5G_5)$. With guanine and cytosine bases stacked alternately, the $d(CG)_8$ oligomer is in the B-form, while $d(G_5C_5)$ is in the A-form. There is Watson–Crick base pairing (Figure 8.7)in these two sequences. According to 2DIR spectroscopy, the coupling between stacked bases varied depending on the secondary structure and the strand sequence. A transition dipole density distribution accurately described the coupling between the bases, indicating that it is primarily electrostatic. These findings suggest that the secondary structure of DNA and RNA affects both IR and 2DIR. A local mode description of amide I carbonyl stretches in peptide backbones was closely associated with the model that some researchers evaluated.

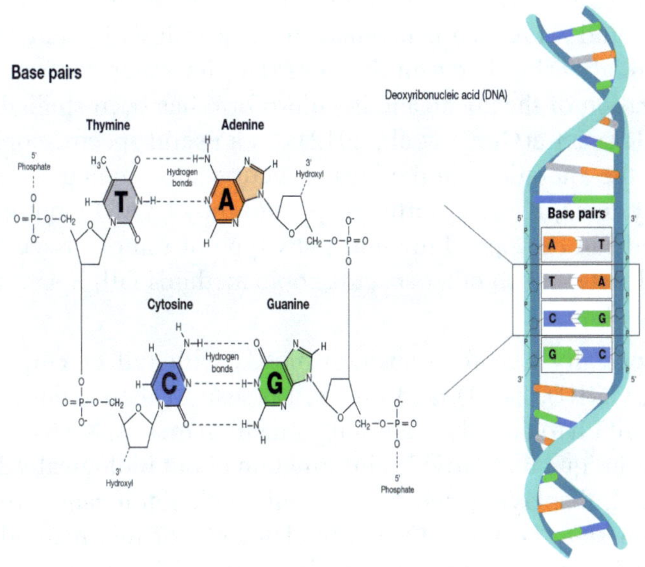

Figure 8.7. Watson – crick base pairing.

Source: https://upload.wikimedia.org/wikipedia/commons/0/05/DNA_base-pair_diagram.jpg, Creative Commons License.

Guanine/cytosine DNA sequences that use vibrational modes localized on specific bases respond well to it. Using finite difference techniques within DFT, most of the coupling terms were accurately replicated. Although the model is promising as is, it currently ignores charge transfer and hydration. Although the model does not yet include the bases adenine and thymine, it has already made it possible to quantitatively understand DNA 2DIR spectra, which was previously impossible. Low water content samples were the subject of research on synthetic double-stranded DNA oligomers with 23 alternating adenine and thymine base pairs. This made it possible to separate the interactions with the aqueous environment from the intra-DNA coupling. Line forms and the various N–H stretching modes were identified. On the adenine and thymine bases, anharmonic couplings between the N–H oscillators were detected. A variety of N–H stretching bands, which are congested and challenging to interpret in FTIR, might be assigned using 2DIR spectroscopy. The adenine NH_2 stretching bands showed time-independent inhomogeneous broadening, a sign of structural instability in the oligomers.

Thymine has a rather uniformly widened N–H stretching band. There is anharmonic connection between the N–H oscillator and the cross peaks that appear in the spectra. They also offer proof of the transfer of vibrational energy. The findings would serve as a benchmark for research on hydrated DNA that examines the relationship between the DNA and its water content. The N–H stretching vibration of the adenosine in chloroform has been studied using the FTIR and 2DIR spectra(Greve et al., 2012). As a useful spectroscopic method for examining the alterations in the local structure surrounding a metal center in biological systems, X-ray absorption spectroscopy (XAS) requires no long-range order and can be applied to almost any type of sample, regardless of its actual physical state. When other spectroscopic methods fail, XAS can provide valuable information.

This is especially true for transition metals with full or empty d-shells, such as Au(I), Ga(III), Zn(II), and others, because of their unique electronic structure. In addition to directly examining transition metals, XAS has proven to be a flexible technique that provides information about biologically significant nonmetals like S, displaying rich XAS signals with significant shifts brought on by oxidation state changes. One special feature of this method is that it allowed for the direct *in situ* assessment of the ratio of glutathione present in its reduced form (GSH) and in its oxidized, disulfide bridged form (GSSG). The average coordination environment and oxidation state of the probed element are measured using bulk XAS techniques over a wider sample area (100 μm^2). The ability to visualize specific areas of tissue or even individual cells and obtain data on the chemical state of the element being studied in various cell compartments is made possible by focused beams. The low concentration of the metal in the targeted tissue itself is the primary barrier to bio-XAS investigations. Synchrotrons and other X-ray sources with photon fluxes between 1×10^{11} and 4×10^{12} photons per second are required to produce spectra with a high signal to noise ratio(Cramer, 2020). Ultra-low temperature sample environments about 20 K can reduce the effects of the high photon flux, which includes the risk of reduction by secondary electrons and rapid sample degradation. The combination of high photon flux, extremely sensitive detectors, and ultra-low sample environments offered by synchrotrons enables the study of almost all biologically significant elements in energy regions ranging from the hard X-ray and gamma-ray to the soft X-ray region. Self-absorption effects brought on by densely packed material and high absorber concentrations frequently impair fluorescence readings. The spectral absorption features are diminished by the produced fluorescence X-rays being reabsorbed in the sample. By comparing an absorption and fluorescence spectrum taken simultaneously, hindering effects in the hard X-ray area can be verified, provided that the sample contains a high enough absorber concentration. Reducing the particle size in the sample or mixing with a low absorbing matrix material, like boron nitride (BN), can both reduce the absorber concentration and prevent or lessen self-absorption. Since the major components of biological matrices are the elements nitrogen, oxygen, hydrogen, and carbon, which have low atomic numbers and absorb primarily in the soft X-ray region, self-absorption is a minor issue when studying transition metals in biological samples that primarily absorb in the hard X-ray region. Reabsorption is reduced in biological samples due to lower concentrations of transition metals and other heavy scatterers like sulfur and chlorine.

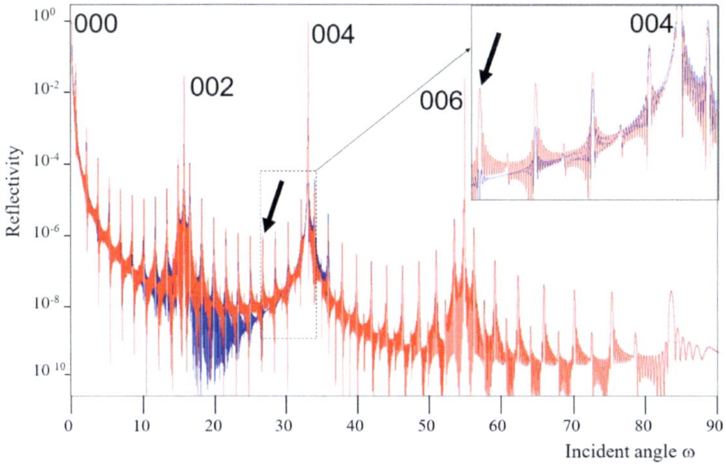

Figure 8.8. Diffraction peaks.

Source: Fewster, P. F. (2023). The limits of x-ray diffraction theory. *Crystals, 13*(3), Article 3. https://doi.org/10.3390/cryst13030521, Creative Commons License.

8.12. DISTORTIONS

Unwanted distortions in the spectra are caused by glitches, diffraction peaks, (Figure 8.8)and pin-hole and thickness effects. Heterogeneous samples with alternating X-ray transparent and opaque sections are susceptible to pin-hole effects. When samples are produced excessively thick, thickness effects are a common issue in transmission measurements. Inhomogeneities, variations in particle size, or errors in beamline instruments can result in diffraction peaks or glitches in the spectra. To reduce the likelihood of pin-hole effects, the samples should be finely ground to reduce the average particle size and the associated interstitial holes. Although particle size is slightly dependent on fluorescence investigations, inhomogeneous samples should be avoided. When a rapid absorption profile is gathered across the entire sample area, inhomogeneity in the sample is detected. By diluting the sample material to an estimated absorption of 1 absorbance unit (AU), thickness effects can be prevented. Biological samples are inherently disordered, having densely packed tissues interspersed with holes or thinner parts. The only method to get beyond the effects in native tissue samples without harming the original tissue matrix is to carefully choose the probed sample location or site. Data capture in transmission mode is appropriate for samples with absorber concentrations larger than 1%, while fluorescence mode is used to measure samples with concentrations between 10 and 100 ppm. The sample should be prepared to enable transmission and fluorescence measurements if the element being studied has a concentration in between. In addition to the harmful effects of X-ray radiation on biological material, sample preparation methods must also be taken into account.

8.13. XAS

XAS measurement samples are frequently prepared in advance in the laboratory. Until all measurements are finished, the sample should be prepared and maintained in a way that ensures its stability during the entire duration. Moisture-sensitive samples can be dried and kept in low humidity environments, whereas samples that are prone to deterioration under oxygen can be kept in sealed containers filled with an inert gas. Biological samples are in immediate risk of biodegradation due to airborne particles, ultraviolet light, and temperature changes. Because of this, biological material is kept in sealed containers at temperatures lower than -80°C in the dark after being flash frozen in liquid nitrogen. The possibility of ice crystal damage is present when biological samples are frozen. A suitable cryoprotectant, such as ethylene glycol or glycerol, should be added if possible. Biomolecule-containing solutions can be created in advance and kept at cryogenic temperatures. Another choice that enables room temperature (RT) storage is freeze drying. It is strongly advised to simultaneously collect multiple appropriate molecules that represent the potential coordination in vivo when doing XAS investigations on biological samples. Only when a suitable reference with a known absolute edge position was measured under the identical experimental conditions can an accurate energy calibration be made. Understanding how ligand exchange reactions affect the general morphologies and energy positions of the spectra, as well as the size of the energy changes connected to oxidation/reduction reactions, is crucial for XANES analysis. Understanding and estimating these effects can be aided by model compound spectra. The model compound structures are becoming increasingly important in the entire analytical process and can be used as preliminary structural models for the theoretical computation of EXAFS and XANES spectra(Rehr and Ankudinov, 2005). The evaluation of edge shape variations, pre- and post-edge features, and comparison with spectra from structurally and chemically related model substances are frequently the main objectives of XANES investigation of biomolecules.

It is necessary to calibrate the edge energy position using a standard compound whose absolute edge energy is known. The edge position is determined over a randomly selected point on the absorption edge, usually the half height of the normalized edge height or the first maximum in the first derivative. There are other accepted definitions for the edge positions, and the edge energy is determined by selecting the maximum of the white line. The optimum spectra for this method are those having a narrow and identifiable white line amplitude. The first unexpected increase in a XAS spectrum is known as the white line, and it denotes electron transitions to unfilled bound states that occur just below the continuum of free electron states. Since transitions into filled orbitals are prohibited due to the Pauli exclusion principle, elements with full d-shells exhibit almost no white line features. The zero crossing of the second derivative is another way to specify the edge position of a XANES spectrum. Some researchers prepared an integrative technique that takes into account changes throughout the entire edge jump, while examining the oxidation states of the various states of the water-splitting manganese complex in photosystem II. The absorption processes and subsequent interactions with nearby atoms in the XANES deviate from the EXAFS approximations. The local electronic arrangement and structure surrounding the absorbing atom have a significant impact on the XANES spectrum. There is currently no widely accepted theoretical explanation for XANES spectrum computation. Over the past few decades, codes like MXAN and FEFF wereaddressing this problem. Two common analysis techniques frequently employed for XANES investigation include statistical methods such as least squares fitting (LSF) of known model compound spectra to unknown spectra and principal component analysis (PCA) in conjunction with target transformation. Protein molecules will accumulate at interfaces when they come into contact with a surface, which can be advantageous or problematic. Therefore, the development of novel synthetic techniques for biomedical coatings and implant devices, biosensors, immunological tests, and drug delivery schemes with enhanced performance could be advanced by efficient and molecular-level control of interfacial protein structures. However, the structure and activity of the interfacial proteins utilized for biological recognition have an impact on a biosensor or biochip performance.

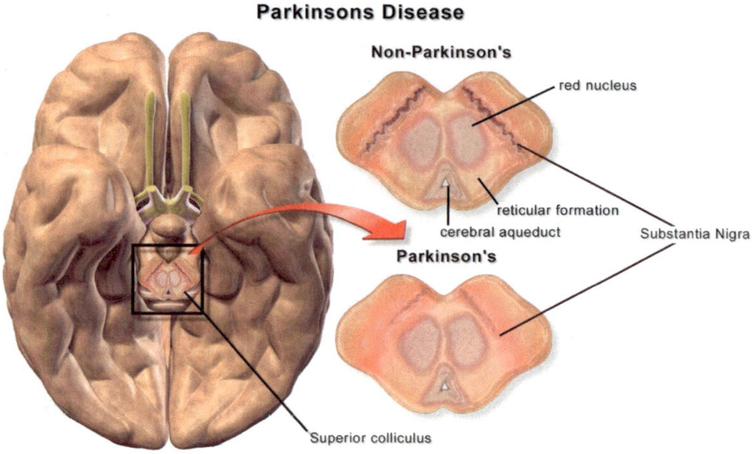

Figure 8.9. Parkinson's disease.

Source: https://upload.wikimedia.org/wikipedia/commons/c/cb/Blausen_0704_ParkinsonsDisease.png, Creative Commons License.

The selectivity and sensitivity of the sensor will be diminished if the protein-binding domains are randomly oriented or deformed; uniformly orienting the molecules for optimal binding is therefore anticipated to improve sensor performance. However, membrane-associated protein structures control biological processes like molecular anchoring, immunological recognition, and signal transmission. Many protein related diseases such as Parkinson's disease(Figure 8.9), Alzheimer's disease(Figure 8.10), and prion disease(Figure 8.11), can be brought on by the aggregation of these membrane proteins in cell membranes. Therefore, to comprehend the nature of interfacial protein interaction, the characteristics of the interfacial biodevices, and disease causes, in situ and real-time characterization of molecular structures and protein orientation at interfaces is crucial. Such research will yield crucial clues for manipulating the biointerface. Using a variety of methods, including atomic force microscopy, nuclear magnetic resonance (NMR), surface plasmon resonance, and others, much research has been conducted to obtain insight into interfacial protein structures and behaviors.

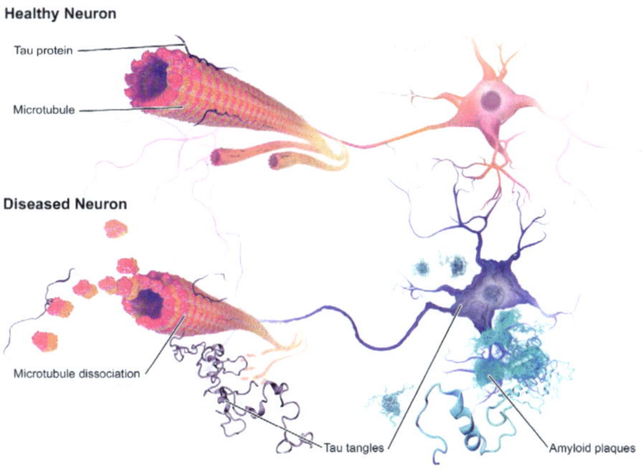

Figure 8.10. Alzheimer's disease.

Source: https://upload.wikimedia.org/wikipedia/commons/9/99/Alzheimers_Disease.jpg, Creative Commons License.

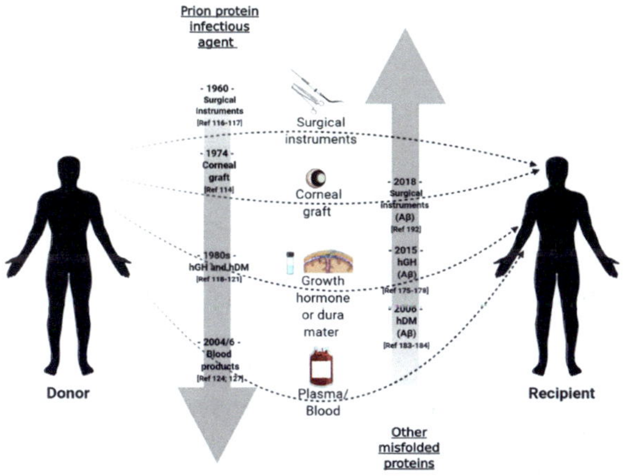

Figure 8.11. Prion disease.

Source: Ritchie, D. L., &Barria, M. A. (2021). Prion diseases: A unique transmissible agent or a model for neurodegenerative diseases? *Biomolecules, 11*(2), Article 2. https://doi.org/10.3390/biom11020207, Creative Commons License.

8.14. SFG–VS

The sum frequency generation vibrational spectroscopy (SFGVS)(Figure 8.12)has grown into a potent and adaptable method for studying molecular-level interfacial molecular structures in situ in various chemical settings(Lu et al., 2017). In addition to identifying interfacial molecular species, SFG-VS, a polarized vibrational spectroscopy, offers insights into the interfacial structures, including the orientation and distribution of functional groups on a surface or at an interface in situ in real time. SFG-VS has been used to model membranes and other interfacial environments and investigate the orientation and structure of different biomolecules, such as proteins and peptides. The interfacial structures of membrane-bounded peptides and proteins, including magainin 2, cecropin P1 (CP1c), melittin, and G protein, have been satisfactorily determined.

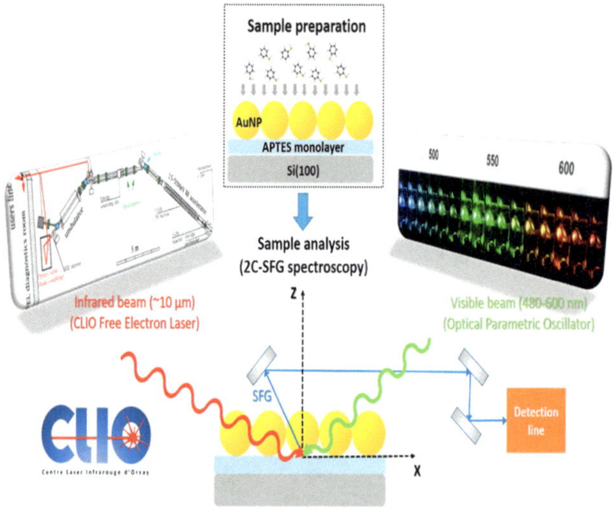

Figure 8.12. Sum frequency generation vibrational spectroscopy.

Source: Humbert, C., Pluchery, O., Lacaze, E., Busson, B., &Tadjeddine, A. (2022). Two-color sum-frequency generation spectroscopy coupled to plasmonicswith the CLIO free electron laser. *Photonics, 9*(2), Article 2. https://doi.org/10.3390/photonics9020055, Creative Commons License.

The secondary structures of the alpha-helix, beta-sheet, and random coil at interfaces were described using SFG-VS. The effectiveness of polyphenol EGCG inhibitors in preventing amyloid formation at phospholipid interfaces was examined. Using SFG, NEXAFS, and NMR spectroscopy, researchers investigated the orientation and conformation of alpha-helix and beta-strand model peptides on various surfaces.

SFG-VS is a second-order nonlinear optical laser method that generates surface and interface vibrational spectra. SFG spectra are now measured using either a broad-band method or a frequency scanning technique. Two picosecond pulse laser beams i.e., visible and infrared, are used in frequency scanning systems, whereas a picosecond narrow-band visible beam and a femtosecond broad-band infrared beam are used in broad-band systems. The IR beam and the pulsed visible beam must overlap on the samples both spatially and temporally in all SFG studies. The nonlinear process produces the SFG signal at the sum frequency of the two input beams. SFG process as a fusion of Raman scattering and IR absorbance. The intensity of the sum frequency signal, as determined by the dipole approximation, is proportional to the intensity of the incoming visible laser beam, IR laser beam, and the square of the second-order nonlinear susceptibility of vibration.

This susceptibility disappears when a material has inversion symmetry. The C–H stretching frequency region was the primary focus of earlier SFG investigations on the structure of interfacial proteins. Utilizing the C–H SFG stretching signals produced by the side chains of gramicidin A, revealed the orientation of gramicidin A in a DMPC monolayer. The signals from protein side chains and other surface molecules make up the SFG signals obtained in the C–H stretching frequency range. Proteins produce backbone vibrations called amide vibrations in addition to side-chain vibrations. Although the amide I vibration is dependent on the secondary structure of the backbone and is not significantly impacted by the side chain composition, all amide frequencies are conformation-sensitive.

The peak centers of various secondary structure types vary. The antiparallel beta-sheet structure has the distinctive peaks at roughly 1630 and 1685 cm^{-1}, but the alpha-helical structure has a peak center at roughly 1655 cm^{-1}. Secondary-structure analysis is where the amide I vibration is frequently employed. They can produce very strong SFG amide I signal of interfacial proteins by using an experimental geometry that is almost reflective. SFG is currently being used by researchers for interfacial protein studies. If the IR transition dipole moment and the Raman polarizability tensor are known, the SFG hyperpolarizability tensor of a vibrational mode can be inferred. Three approaches can be used to determine the IR transition dipole moment and Raman polarizability tensor: ab initio computation, bond additivity model, and polarized IR and Raman spectroscopy techniques. ab initio calculations were used to determine the IR transition dipole moment and Raman polarizability tensor utilizing tripeptide segments as modeling peptides. However, because proteins are large molecules, the calculation becomes challenging and complex. Using the bond additivity model in accordance with the symmetry and structure, the Raman polarizability tensor and IR transition dipole moment of a peptide unit can be used to determine the Raman polarizability tensor and IR transition dipole moment for a secondary structure. The experimentally measured values obtained by polarized Raman and IR spectra measurements are compared to and validated by the predicted values inferred by the bond additivity model. When working with large molecules like proteins and polymers, the bond additivity model exhibits more benefits. Proteins and DNA are examples of biomolecules that are essential to the biological processes of living cells.

8.15. BIOMOLECULES

Biomolecules are typically studied experimentally in diluted aqueous solutions. But because biomolecules function in cells, which are packed with diverse types of molecules, their environments differ from those of their diluted aqueous solution. The biomolecular function in living cells is impacted by macromolecular congestion. The biomolecular function will be impacted by variations in the condition of the surrounding fluids. The aqueous cavity radius and the amount of water molecules within the reverse micelle are two examples of structural parameters of protein-free reverse micelles that have been established. The radius of the aqueous cavity can be manipulated experimentally and is proportional to the water–surfactant molar ratio (W_0). One of the important parameters for knowing the structural properties of reverse micelles is the W_0 value. While it has been shown that the characteristics of the protein in reverse micelles depend on W_0, determining the structural parameters of PCR micelles as a function of W_0 has proven to be a difficult problem. It should be noted that the binary system known as PCR micellar solution is made up of both PCR micelles and protein-unfilled ones, or water-containing reverse (WCR) micelles. Understanding the structural characteristics of PCR micelles is essential for a detailed investigation of the protein and water within reverse micelles.

8.16. MICELLE

W_0 also represents the water-to-surfactant molecule ratio of the reverse micelle. The W_0 value of the PCR micelle and the PCR micellar solution are same. To create a model to know the structural properties of PCR micelles, this problem must be fixed. We concentrate on the water O–H stretching vibration band in order to do this. A commercial spectrophotometer measuring the O–H stretching vibration of water in WCR micellar solution at room temperature displays the W_0 dependency of the NIR absorption spectrum(Murakami, 2013). AOT and isooctane are utilized as surfactants and solvents, respectively, for the reverse micelles for example. The spectral characteristic has a prominent dependence on W_0. The environment of the water molecule, the hydrogen bond state, affects the O–H stretching vibration. The core-shell model explains the spectrum W_0 dependence. According to the model, WCR micelles include two different kinds of water: interfacial water that is bonded to the surfactant and bulk-like water that is situated in the center of the aqueous cavity. As micelle size, or W_0, increases, so does the ratio of bulk-like to interfacial waters. As W_0 increases, the absorption spectrum moves closer to that of bulk water. Therefore, the W_0 of WCR micelles can be determined from the absorption spectra of the O–H stretching vibration of water. Understanding the catalysis of enzymes requires high-resolution characterization of the active site interactions. The enzymes lower the reaction barrier by stabilizing the transition state structure and disrupting the ground state structure upon binding a substrate. This process gives the enzymes their high catalytic ability. The specificity of an enzyme is also determined by the interactions that occur between it and its substrate both in the ground state and in the transition state. The main causes of the substrate distortions upon binding to the enzyme are frequently the active site noncovalent interactions, such as hydrogen bonding and ionic interactions, which move the substrate closer to the transition state structurally and facilitate the bond-breaking/bond-making process. During the reaction, proton transfer from one functional group to the other is frequently facilitated by active site interactions. Rational drug design researchers may benefit from knowledge of active site proton movements and substrate bond distortions, which serve as the foundation for a deeper understanding of enzyme catalysis. Conducting steady state investigations on stable enzyme substrate complexes or enzyme inhibitor complexes that mimic Michaelis complexes or transition states along the reaction coordinate is one method to understand how the distortions on the functional groups of the substrate progress in the enzyme catalysis. Enzyme active site contacts are frequently quantitatively determined by studies that compare the vibrational frequencies or NMR chemical shifts of enzyme-bound ligands to their solution values. These methods can detect minute bond order changes of bonds that an enzyme generates on its substrate, as well as small ligand conformational changes that may be easily missed (ignored) in X-ray structures.

Their spatial resolutions are higher than those of X-ray crystallography (Figure 8.13). To track the proton positions in the enzyme complexes, these instruments are also perfectly suited for determining the ionic states and tautomeric forms of substrates and inhibitors. Comparing the chemical shifts or vibrational frequencies of the ligand molecule acquired in solution and in the enzyme–ligand complex is usually a better way to get qualitative information about the ligand structure and interaction.

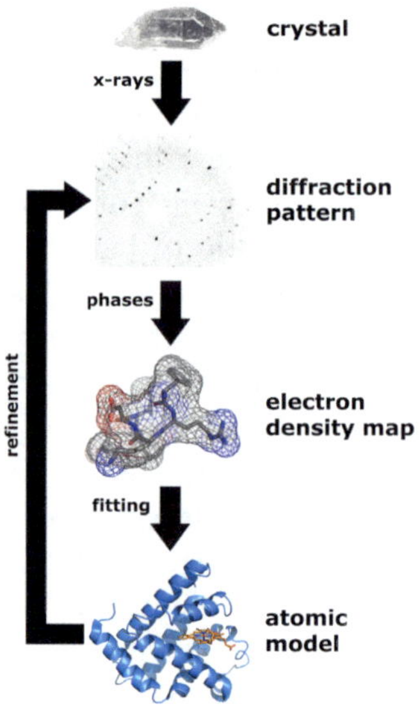

Figure 8.13. X-ray crystallography.

Source: https://upload.wikimedia.org/wikipedia/commons/7/73/X_ray_diffraction.png, Creative Commons License.

Isotope-edited difference vibrational spectroscopy or isotope-edited NMR techniques can be used to solve the spectral congestion or crowding issue when the enzyme is present. Empirical correlations and computer simulations based on ab initio techniques may yield more quantitative data, such as ligand conformational changes, ligand bond length changes, and the hydrogen bonding energy changes that occur to some groups upon ligand binding to enzymes. There are two steps involved in determining the vibrational structure of molecular groups in proteins: identify the vibrational modes of interest that every other protein band obscures. Researchers created isotope-edited difference Raman and FTIR spectroscopy for this purpose. Using empirical correlations or ab initio normal mode calculations, we can derive structural information about the chemical groups from the observed vibrational modes, particularly their modifications upon protein interaction. We can utilize Raman difference spectroscopy, which has been applied to a variety of enzyme systems, to capture particular structural features of the LDH-bound NADH. This technique obtains the vibrational spectrum of the bound ligand by subtracting the apoprotein spectrum from that of the complex. Isotope editing expands the use of Raman difference spectroscopy to systems where apoprotein instability or excessive protein conformational changes caused by the bound ligand that manifest in the difference spectrum make apoprotein spectrum measurement impractical. Two protein–ligand complexes, one of which is selectively tagged with a stable isotope, generate an isotope-edited difference spectrum. In the difference spectrum, vibrational modes linked to the isotopic tag manifest as spectral shifts and positive–negative band pairings, but all other bands cancel.

The C–D stretch frequencies of the LDH-bound NADH have been established for LDH/NADH and LDH/NADH/oxamate complexes using NADH that has been selectively tagged at the A side (pro-R) or B side (pro-S) of the reduced nicotinamide with deuterium. The C_4–D stretch frequencies of the reduced nicotinamide ring undergo two significant modifications upon NADH binding to LDH: the frequency difference between the two C_4–D stretch modes increase by 32 cm^{-1}, and the average of the two C_4–D frequencies increase by 14 cm^{-1}. Employing pro-R or pro-S side deuterium labeling, two resolved C_4–D bands are seen upon oxamate binding. In both situations, the intensity ratios of these two bands are roughly 2:1. Researchers have used normal mode analysis based on quantum mechanical ab initio and semiempirical calculations to quantitatively interpret these spectral shifts in terms of nicotinamide structural alterations. The two resolved bands in each of the two C_4–D labeled NADH spectra demonstrate that the reduced nicotinamide ring in the LDH/NADH/oxamate ternary complex adopts two conformations. The main conformer resembles the conformer in the binary complex when the pro-S deuteron is in pseudoequatorial geometry and the pro-R C_4 deuteron is brought to a pseudoaxial. The pro-R C_4 deuteron is pseudoequatorial and the pro-S deuteron is pseudoaxial in the minor conformer, which makes up around half of the major population. The C_4–D stretch bands have narrowed from the binary complexes in all cases, suggesting that the ternary complex has greatly reduced ring puckering processes. In comparison to the decreased nicotinamide $C_2 = C_3$ bond, the carboxylamide C = O bond is transoid oriented. A quantitative relationship between hydrogen bonding energy and C = O stretch frequency can be ascertained since the C = O stretch mode is fairly prominent in Raman or IR spectra. It is therefore perfectly suited for research on the activation of substrate C–O bonds in enzyme complexes.

8.17. BONDING

Hydrogen bonding energy vs NH/OH chemical shift or C = O stretch frequency correlations have also been developed with the help of ab initio calculations. Some studies discovered a relationship between the shift of the C = O stretching frequency and the bond energy between a simple molecule C = O bond and its hydrogen bonding/ionic interaction counterparts. Two infrared bands were observed in the complex for $^{13}C_2$ labeled oxamate: a major band at 1,606 cm^{-1} and a minor band at 1,623 cm^{-1}. At 1,643 and 1,656 cm^{-1}, respectively, they are easily recognized. This suggests that there are two distinct types of oxamate in the complex, with a population ratio of roughly 2:1. The 14 cm^{-1} red shift of its C2 = O stretch frequency from 1,620 cm^{-1} in solution to 1,606 cm^{-1} in the complex indicates that the oxamate C2 = O bond undergoes a stronger hydrogen bonding in the major form than in solution. The 3 cm^{-1} blue shift of the C2 = O stretch frequency, from 1,620 cm^{-1} in solution to 1,623 cm^{-1} in the complex, indicates that the hydrogen bond to the oxamate C2 = O bond in the minor form is comparable to weaker than that in solution. One distinction between these two types of bound oxamate and the solution spectra is that the band width at half maximum drops from 30 to 6–7 cm^{-1}, suggesting a clearly defined structure and environment. The standard for analyzing the atomistic features of three-dimensional structures of proteins complexed with small ligands is nuclear magnetic resonance (NMR) and X-ray diffraction. The structures of proteins, DNA, and RNA have been effectively studied using these approaches. NMR and X-ray have also been used to investigate the interactions between biomolecules, specifically the binding of small ligands to proteins, due to the development of advanced instruments and improved methodologies like NOESY, crystal soaking, COSY,etc. Despite its strength, X-ray and NMR both have certain inherent drawbacks. Both need high doses, usually in the mM range. Extreme conditions are needed to form protein crystals for X-ray diffraction, which frequently raises concerns about how accurately the X-ray structures replicate the solution structure. NMR necessitates the selective isotopic labeling of the molecules under investigation, which frequently calls for the employment of ad-hoc recombinant proteins to integrate the isotopes at the appropriate sites.

8.18. APPROACHES

Alternative approaches and procedures are required to investigate the interaction of proteins with small ligands because of the sample preparation considerations as well as others. The study of proteins and protein/ligand binding heavily relies on optical spectroscopy and optical methods. To investigate protein–protein, protein–DNA, and protein–small ligand binding, absorption and fluorescence spectroscopy including some of its more complex uses like fluorescence resonance energy transfer, or FRET remains a popular option. Optical spectroscopy is unable to provide the resolution required to obtain direct atomistic details of the interaction because the incident radiation used in these experiments has a wavelength larger than the molecules under investigation. Retrieving better resolution data and even three-dimensional details may become possible by combining one or more optical spectroscopies with computer techniques. Samples requiring optically thin conditions i.e., low concentrations are among those for which NMR or X-ray do not provide a practical solution. The studies of ligand-mediated photoinduced effects on proteins offer an example of such samples. The challenges of such problems must be addressed using lower resolution techniques, such as identifying the binding location of the photoactive ligand on the protein, characterizing the structural changes of the polypeptides before and after irradiation, determining the mechanisms that trigger the conformational changes, etc. These studies not only require micromolar concentration, but they also rely on comparing structures before and after or during irradiation with visible light, which makes NMR or X-ray methods inappropriate. Indirect information can be obtained for this purpose using a variety of methods.

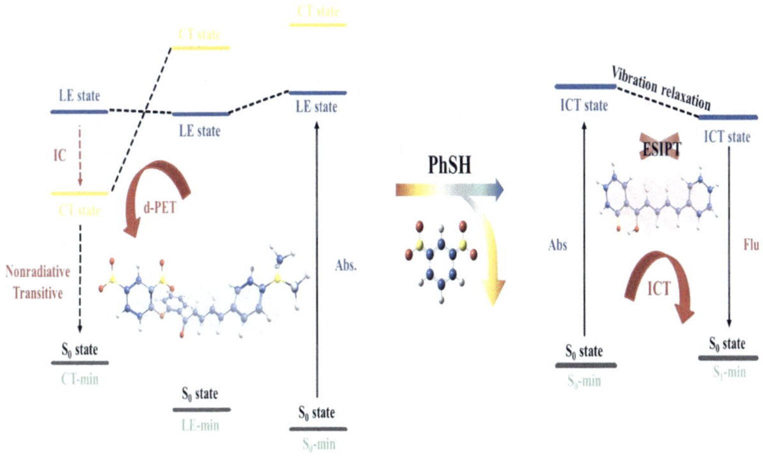

Figure 8.14. Photoinduced charge transfer.

Source: Wang, Y., Zhang, M., Li, W., Wang, Y., & Zhou, P. (2023). Theoretical investigation on the "ON-OFF"mechanism of a fluorescent probe for thiophenols: photoinduced electron transfer and intramolecular charge transfer. *Molecules, 28*(19), Article 19. https://doi.org/10.3390/molecules28196921, Creative Commons License.

By analyzing the variations in absorption and emission of the ligand and/or the protein, absorption and fluorescence spectroscopy can be used to ascertain whether the ligand docks the proteins and to recover binding properties. Information regarding the ligand variations between its free and docked states can also be obtained via resonance Raman spectroscopy (RRS). Details regarding the interaction between a ligand and a protein can also be ascertained by nanosecond and sub-nanosecond fluorescence decay. Small molecules that, when excited by visible light, trigger different photophysical relaxation processes that may impact the surrounding environment and, specifically, the structure of the protein they are docked to, constitute a class of significant protein ligands. A few examples of photoactive ligands are pteridines and chlorophylls. Photoinduced processes, in which photoactive ligands bind to proteins are widely used.

8.19. PHOTOINDUCED CHARGE TRANSFER (PCT) AND PET

PCT(Figure 8.14)is likely the common use of photoactive processes. Since photoinduced electron transfer (PET) serves as the catalyst for photosynthesis, which transforms photon energy into chemical energy that can be stored, it is likely the significant event for maintaining life on Earth. Certain microorganisms and plants rely on this mechanism to survive. Higher species have chosen PET for different purposes. The utilization of the spin of the electron produced in PET by certain migratory birds to position themselves in the magnetic field during the day is an interesting example. However, nature uses other photoinduced events besides PCT. In order to initiate the visual response, specialized molecules (retinal) integrated transmembrane protein receptor need photon energy to undergo a cis–trans isomerization. This process is known as photoisomerization. Proteins complexed with the photoactive ligands undergo conformational changes as a result of the photophysical event. The polypeptide conformational changes then activate additional proteins, or receptors, which sets off a signaling cascade that results in the biological reaction. Photon energy is converted into electrical potential energy across a membrane in the photosynthetic mechanism. This energy is then utilized to convert NADP into NADPH and simultaneously make ATP and carbohydrates. This method relies on the ability to maintain charge separation across the membrane, temporarily. This is accomplished in photosynthesis by a series of molecules, but the fundamental process is the induction of protein conformational changes. These conformational changes enable protons to separate across the membrane and create an electric potential that serves as a source of free energy for the synthesis of chemical energy, making charge recombination slower than charge separation. Ferrodoxin-$NADP^+$ reductase serves as an illustration of one of the conformational changes throughout the photosynthetic pathway, where the reduction in $NADP^+$ is based on a localized conformational change of the enzyme.

The transmembrane proteins (rhodopsins) undergo photoisomerization of the photoactive ligand, which causes the proteins to shift shape. Such a structural alteration can either encourage binding to G-proteins, which sets off a signaling cascade that depolarizes the neuronal membrane and sends the stimulus to the optical nerve, or it can cause the transmembrane channel to open, which would depolarize the cellular membrane and stimulate the optical nerve. A protein's structural change in response to the photoinduced event is a crucial step in converting photon energy into a biological reaction. In certain instances, a biological effect on photon energy conversion is obtained by combining many photophysical phenomena. Several bacteria employ a combination of photoisomerization for a number of purposes. The protein conformational change that binds the photoactive ligand determines the biological impact of the photoinduced process. The photoactive yellow protein found in the halophile *Ectothiorhodospira* is one example. When a photon is absorbed, the pigment molecule undergoes photoisomerization, which is followed by proton transfer from a glutamic acid residue and a significant but reversible conformational change. This leads to a signaling cascade that has a variety of biological effects. Researchers have looked into nature-mimicking models of dye/protein complexes because of the fundamental role that photophysical and photochemical events play in biological functions. This has been done either to understand how certain events produce their biological effects or to take advantage of the photoinduced functions for other purposes.

BIBLIOGRAPHY

1. Aballe, L., Foerster, M., Cabrejo, M., Prat, J., Pittana, P., Sergo, R., Lucian, M., Barnaba, M., Menteş, T. O., & Locatelli, A. (2019). Pulse picking in synchrotron-based XPEEM. *Ultramicroscopy, 202*, 10–17. https://doi.org/10.1016/j.ultramic.2019.03.011 (accessed on 30th July 2025).
2. Ahluwalia, V. K. (2023a). Infrared spectroscopy. In V. K. Ahluwalia (Ed.), *Instrumental Methods of Chemical Analysis* (pp. 179–231). Springer Nature Switzerland. https://doi.org/10.1007/978-3-031-38355-7_23 (accessed on 30th July 2025).
3. Ahluwalia, V. K. (2023b). Infrared spectroscopy. In V. K. Ahluwalia (Ed.), *Instrumental Methods of Chemical Analysis* (pp. 179–231). Springer Nature Switzerland. https://doi.org/10.1007/978-3-031-38355-7_23 (accessed on 30th July 2025).
4. Ahmad, A., &Ayub, H. (2022). Fourier transform infrared spectroscopy (FTIR) technique for food analysis and authentication. In P. B. Pathare& M. S. Rahman (Eds.), *Nondestructive Quality Assessment Techniques for Fresh Fruits and Vegetables* (pp. 103–142). Springer Nature. https://doi.org/10.1007/978-981-19-5422-1_6 (accessed on 30th July 2025).
5. Alday, J., Mazzeo, A., & Suarez, S. (2020). Selective detection of gasotransmitters using fluorescent probes based on transition metal complexes. *InorganicaChimica Acta, 510*, 119696. https://doi.org/10.1016/j.ica.2020.119696 (accessed on 30th July 2025).
6. Aly, A. A., Abdallah, E. M., Ahmed, S. A., Rabee, M. M., &Bräse, S. (2023). Transition metal complexes of thiosemicarbazides, thiocarbohydrazides, and their corresponding carbazones with Cu(I), Cu(II), Co(II), Ni(II), Pd(II), and Ag(I)—A review. *Molecules, 28*(4), Article 4. https://doi.org/10.3390/molecules28041808 (accessed on 30th July 2025).

7. Azeez, S., &Shenbagaraman, R. (2025). 8—Fourier transform infrared spectroscopy in characterization of bionanocomposites. In S. Ahmed & C. M. Hussain (Eds.), *Characterization Techniques in Bio Nanocomposites* (pp. 209–227). Woodhead Publishing. https://doi.org/10.1016/B978-0-443-22067-8.00008-3 (accessed on 30th July 2025).

8. Baiz, C. R., McRobbie, P. L., Anna, J. M., Geva, E., &Kubarych, K. J. (2009). Two-dimensional infrared spectroscopy of metal carbonyls. *Accounts of Chemical Research, 42*(9), 1395–1404. https://doi.org/10.1021/ar9000263 (accessed on 30th July 2025).

9. Bakhmutov, V. I. (2015). *NMR Spectroscopy in Liquids and Solids*. CRC Press. https://doi.org/10.1201/b18341 (accessed on 30th July 2025).

10. Balinova, Yu. A., Varrik, N. M., Istomin, A. V., &Lyulyukina, G. Yu. (2018). Zirconium dioxide fibers and methods of obtaining them. *Fiber Chemistry, 50*(1), 10–18. https://doi.org/10.1007/s10692-018-9919-x (accessed on 30th July 2025).

11. Barbillat, J., & da Silva, E. (1997). Near infra-red Raman spectroscopy with dispersive instruments and multichannel detection. *Spectrochimica Acta Part A: Molecular and Biomolecular Spectroscopy, 53*(13), 2411–2422. https://doi.org/10.1016/S1386-1425(97)00181-9 (accessed on 30th July 2025).

12. Barnes, F., Polk, C., &Greenebaum, B. (2018). 0: Introduction to electromagnetic fields. In *Bioengineering and Biophysical Aspects of Electromagnetic Fields, Fourth Edition* (4th ed.). CRC Press.

13. Barr, T. L. (1991). Advances in the application of X-ray photoelectron spectroscopy (ESCA) Part I. Foundation and established methods. *Critical Reviews in Analytical Chemistry, 22*(1, 2), 567–635. https://doi.org/10.1080/10408349108055027 (accessed on 30th July 2025).

14. Baryshnikov, G., Minaev, B., &Ågren, H. (2017). Theory and calculation of the phosphorescence phenomenon. *Chemical Reviews, 117*(9), 6500–6537. https://doi.org/10.1021/acs.chemrev.7b00060 (accessed on 30th July 2025).

15. Bifulco, G., Dambruoso, P., Gomez-Paloma, L., & Riccio, R. (2007). Determination of relative configuration in organic compounds by NMR spectroscopy and computational methods. *Chemical Reviews, 107*(9), 3744–3779. https://doi.org/10.1021/cr030733c (accessed on 30th July 2025).

16. Bitarafan, M. H., &DeCorby, R. G. (2017). On-chip high-finesse Fabry–Perot microcavities for optical sensing and quantum information. *Sensors, 17*(8), Article 8. https://doi.org/10.3390/s17081748 (accessed on 30th July 2025).

17. Björn, L. O. (2015). The nature of light and its interaction with matter. In L. O. Björn (Ed.), *Photobiology: The Science of Light and Life* (pp. 1–20). Springer. https://doi.org/10.1007/978-1-4939-1468-5_1 (accessed on 30th July 2025).

18. Blundell, S. J. (1999). Spin-polarized muons in condensed matter physics. *Contemporary Physics, 40*(3), 175–192. https://doi.org/10.1080/001075199181521 (accessed on 30th July 2025).

19. Borjian, N., Farhadian, S., Shareghi, B., Asgharzadeh, S., Momeni, L., & Ghobadi, S. (2025). Binding affinity and mechanism of dicofol–lysozyme interaction: Insights from multi-spectroscopy and molecular dynamic simulations. *International Journal of Biological Macromolecules, 308*, 142569. https://doi.org/10.1016/j.ijbiomac.2025.142569 (accessed on 30th July 2025).

20. Böttcher, R., &Dziesiaty, J. (1972). Ligand hyperfine interaction of Ti^{2+} in CdS and CdSe on sites with trigonal symmetry $C_{3}v$. *Physica Status Solidi (b), 53*(2), 505–518. https://doi.org/10.1002/pssb.2220530212 (accessed on 30th July 2025).

21. Boulos, M. I., Fauchais, P. L., &Pfender, E. (2023). Electrode phenomena in plasma sources. In M. I. Boulos, P. L. Fauchais, & E. Pfender (Eds.), *Handbook of Thermal Plasmas* (pp. 553–594). Springer International Publishing. https://doi.org/10.1007/978-3-030-84936-8_13 (accessed on 30th July 2025).

22. Brittain, H. G. (2006). Electromagnetic radiation and spectroscopy. In *Spectroscopy of Pharmaceutical Solids*. CRC Press.

23. Bryant, R. G. (1978). NMR relaxation studies of solute–solvent interactions. *Annual Review of Physical Chemistry, 29*, 167–188. https://doi.org/10.1146/annurev.pc.29.100178.001123 (accessed on 30th July 2025).

24. Campanella, B., Palleschi, V., &Legnaioli, S. (2021). Introduction to vibrational spectroscopies. *ChemTexts, 7*(1), 5. https://doi.org/10.1007/s40828-020-00129-4 (accessed on 30th July 2025).

25. Cao, H., Chriki, R., Bittner, S., Friesem, A. A., & Davidson, N. (2019). Complex lasers with controllable coherence. *Nature Reviews Physics, 1*(2), 156–168. https://doi.org/10.1038/s42254-018-0010-6 (accessed on 30th July 2025).

26. Cazaux, J. (2010). Secondary electron emission and charging mechanisms in Auger electron spectroscopy and related e-beam techniques. *Journal of Electron Spectroscopy and Related Phenomena, 176*(1), 58–79. https://doi.org/10.1016/j.elspec.2009.06.004 (accessed on 30th July 2025).

27. Chen, X., Sun, Y., Xiu, Z., Li, X., & Zhang, D. (2006). Stoichiometric analysis of biological hydrogen production by fermentative bacteria. *International Journal of Hydrogen Energy, 31*(4), 539–549. https://doi.org/10.1016/j.ijhydene.2005.03.013 (accessed on 30th July 2025).

28. Chen, X., Wang, N., He, C., & Lin, X. (2023). Development of all-fiber nanosecond oscillator using actively Q-switched technologies and modulators. *Optics & Laser Technology, 157*, 108709. https://doi.org/10.1016/j.optlastec.2022.108709 (accessed on 30th July 2025).

29. Chifotides, H. T., Giles, I. D., & Dunbar, K. R. (2013). Supramolecular architectures with π-acidic 3,6-bis(2-pyridyl)-1,2,4,5-tetrazine cavities: Role of anion–π interactions in the remarkable stability of Fe(II) metallacycles in solution. *Journal of the American Chemical Society, 135*(8), 3039–3055. https://doi.org/10.1021/ja3082473 (accessed on 30th July 2025).

30. Cialla, D., März, A., Böhme, R., Theil, F., Weber, K., Schmitt, M., & Popp, J. (2012). Surface-enhanced Raman spectroscopy (SERS): Progress and trends. *Analytical and Bioanalytical Chemistry, 403*(1), 27–54. https://doi.org/10.1007/s00216-011-5631-x (accessed on 30th July 2025).

31. Cramer, S. P. (2020). Synchrotron radiation fundamentals. In S. P. Cramer (Ed.), *X-Ray Spectroscopy with Synchrotron Radiation: Fundamentals and Applications* (pp. 39–68). Springer International Publishing. https://doi.org/10.1007/978-3-030-28551-7_3 (accessed on 30th July 2025).

32. da Silva Sousa, J., do Nascimento, H. O., de Oliveira Gomes, H., & do Nascimento, R. F. (2021). Pesticide residues in groundwater and surface water: Recent advances in solid-phase extraction and solid-phase microextraction sample preparation methods for multiclass analysis by gas chromatography-mass spectrometry. *Microchemical Journal, 168*, 106359. https://doi.org/10.1016/j.microc.2021.106359 (accessed on 30th July 2025).

33. Dadarlat, V. M., & Saxena, V. K. (1999). The effect of structural water molecules on the normal mode spectrum of dTn·dAn × dTn DNA. *Journal of Biomolecular Structure and Dynamics, 17*(2), 333–346. https://doi.org/10.1080/07391102.1999.10508365 (accessed on 30th July 2025).

34. De Maleissye, J. T., Lempereur, F., Marsal, C., & Ben-Aïm, R. I. (1976). Infrared photosensitized decomposition of ethane in a cw CO_2 laser beam. *Chemical Physics Letters, 42*(1), 46–49. https://doi.org/10.1016/0009-2614(76)80548-9 (accessed on 30th July 2025).

35. Delsuc, M. A., & O'Connor, P. (2024). The Fourier transforms in analytical science. *Nature Reviews Methods Primers, 4*(1), 1–16. https://

doi.org/10.1038/s43586-024-00326-2 (accessed on 30th July 2025).

36. Demtröder, W. (2010). Experimental techniques in atomic and molecular physics. In W. Demtröder (Ed.), *Atoms, Molecules and Photons: An Introduction to Atomic-, Molecular- and Quantum Physics* (pp. 421–486). Springer. https://doi.org/10.1007/978-3-642-10298-1_11 (accessed on 30th July 2025).
37. Demtröder, W. (2015). Laser spectroscopy in molecular beams. In W. Demtröder (Ed.), *Laser Spectroscopy 2: Experimental Techniques* (pp. 183–224). Springer. https://doi.org/10.1007/978-3-662-44641-6_4 (accessed on 30th July 2025).
38. Efimov, A. M., &Pogareva, V. G. (2006). IR absorption spectra of vitreous silica and silicate glasses: The nature of bands in the 1300 to 5000 cm^{-1} region. *Chemical Geology, 229*(1), 198–217. https://doi.org/10.1016/j.chemgeo.2006.01.022 (accessed on 30th July 2025).
39. El Ghazali, A., Aboulouard, A., Gultekin, B., Tounsi, A., & El Idrissi, M. (2023). Theoretical investigation of novel electron donors for bulk heterojunction solar cells with potential photovoltaic characteristics. *Journal of Molecular Graphics and Modelling, 125*, 108622. https://doi.org/10.1016/j.jmgm.2023.108622 (accessed on 30th July 2025).
40. El-Sayed, M. A. (2002). Triplet state. Its radiative and nonradiative properties (world) [Research-article]. *ACS Publications; American Chemical Society*. https://doi.org/10.1021/ar50001a002 (accessed on 30th July 2025).
41. Ershkov, M. N., Solokhin, S. A., &Smetanin, S. N. (2015). Self-Q-switching of a loop laser cavity with a self-pumped four-wave phase-conjugate mirror in an active laser medium. *Optics and Spectroscopy, 119*(3), 526–533. https://doi.org/10.1134/S0030400X15090088 (accessed on 30th July 2025).
42. Famiglini, G., Palma, P., Termopoli, V., &Cappiello, A. (2021). The history of electron ionization in LC-MS, from the early days to modern technologies: A review. *Analytica Chimica Acta, 1167*, 338350. https://doi.org/10.1016/j.aca.2021.338350 (accessed on 30th July 2025).
43. Feng, X., Wang, X., Redshaw, C., & Tang, B. Z. (2023). Aggregation behavior of pyrene-based luminescent materials, from molecular design and optical properties to application. *Chemical Society Reviews, 52*(19), 6715–6753. https://doi.org/10.1039/D3CS00251A (accessed on 30th July 2025).
44. Fleming, D. G., Cottrell, S. P., McKenzie, I., &Ghandi, K. (2015). Rate constants for the slow Mu + propane abstraction reaction at 300 K by

diamagnetic RF resonance. *Physical Chemistry Chemical Physics, 17*(30), 19901–19910. https://doi.org/10.1039/C5CP02576A (accessed on 30th July 2025).

45. Fleming, D. G., Cottrell, S. P., McKenzie, I., & Macrae, R. M. (2012). New results for the formation of a muoniated radical in the Mu + Br_2 system: A van der Waals complex or evidence for vibrational bonding in Br–Mu–Br? *Physical Chemistry Chemical Physics, 14*(31), 10953–10966. https://doi.org/10.1039/C2CP41366C (accessed on 30th July 2025).

46. Fleming, I., & Williams, D. (2019). Infrared and Raman spectra. In I. Fleming & D. Williams (Eds.), *Spectroscopic Methods in Organic Chemistry* (pp. 85–121). Springer International Publishing. https://doi.org/10.1007/978-3-030-18252-6_3 (accessed on 30th July 2025).

47. Flores-Mancera, M. A., Villarrubia, J. S., & Massillon-JL, G. (2020). Electron inelastic mean free paths for LiF, CaF_2, Al_2O_3, and liquid water from 433 keV down to the energy gap. *ACS Omega, 5*(8), 4139–4147. https://doi.org/10.1021/acsomega.9b03872 (accessed on 30th July 2025).

48. Fujita, N., Jones, R., Öberg, S., &Briddon, P. R. (2007). First-principles study on the local vibrational modes of nitrogen–oxygen defects *in silicon. Physica B: Condensed Matter, 401, 402*, 159–162. https://doi.org/10.1016/j.physb.2007.08.136 (accessed on 30th July 2025).

49. Fultz, B. (2010). Vibrational thermodynamics of materials. *Progress in Materials Science, 55*(4), 247–352. https://doi.org/10.1016/j.pmatsci.2009.05.002 (accessed on 30th July 2025).

50. García-González, D. L., Baeten, V., Pierna, J. A. F., & Tena, N. (2013). Infrared, Raman, and fluorescence spectroscopies: Methodologies and applications. In R. Aparicio & J. Harwood (Eds.), *Handbook of Olive Oil: Analysis and Properties* (pp. 335–393). Springer US. https://doi.org/10.1007/978-1-4614-7777-8_10 (accessed on 30th July 2025).

51. Gel'mukhanov, F., &Ågren, H. (1999). Resonant X-ray Raman scattering. *Physics Reports, 312*(3), 87–330. https://doi.org/10.1016/S0370-1573(99)00003-4 (accessed on 30th July 2025).

52. Gerig, J. T. (2008). Chapter 2 NOE studies of solvent–solute interactions. In G. A. Webb (Ed.), *Annual Reports on NMR Spectroscopy* (Vol. 64, pp. 21–215). Academic Press. https://doi.org/10.1016/S0066-4103(08)00002-1 (accessed on 30th July 2025).

53. Giliberti, M., &Lovisetti, L. (2024). The photoelectric effect and the electron charge. In M. Giliberti& L. Lovisetti (Eds.), *Old Quantum Theory and Early Quantum Mechanics: A Historical Perspective Commented for The Inquiring Reader* (pp. 159–228). Springer Nature Switzerland. https://

doi.org/10.1007/978-3-031-57934-9_5 (accessed on 30th July 2025).

54. Giridhar, G., Manepalli, R. R. K. N., &Apparao, G. (2017). Chapter 7—Confocal Raman spectroscopy. In S. Thomas, R. Thomas, A. K. Zachariah, & R. K. Mishra (Eds.), *Spectroscopic Methods for Nanomaterials Characterization* (pp. 141–161). Elsevier. https://doi.org/10.1016/B978-0-323-46140-5.00007-8 (accessed on 30th July 2025).
55. Glasbeek, M., & Zhang, H. (2004). Femtosecond studies of solvation and intramolecular configurational dynamics of fluorophores in liquid solution. *Chemical Reviews, 104*(4), 1929–1954. https://doi.org/10.1021/cr0206723 (accessed on 30th July 2025).
56. Goldstein, J. I., Newbury, D. E., Michael, J. R., Ritchie, N. W. M., Scott, J. H. J., & Joy, D. C. (2018). X-rays. In J. I. Goldstein, D. E. Newbury, J. R. Michael, N. W. M. Ritchie, J. H. J. Scott, & D. C. Joy (Eds.), *Scanning Electron Microscopy and X-ray Microanalysis* (pp. 39–63). Springer. https://doi.org/10.1007/978-1-4939-6676-9_4 (accessed on 30th July 2025).
57. Goltsev, V., Zaharieva, I., Chernev, P., & Strasser, R. J. (2009). Delayed fluorescence in photosynthesis. *Photosynthesis Research, 101*(2), 217–232. https://doi.org/10.1007/s11120-009-9451-1 (accessed on 30th July 2025).
58. Goodfellow, R. C. (1986). Semiconductor materials and structures for optical-communication devices. In P. N. Butcher, N. H. March, & M. P. Tosi (Eds.), *Crystalline Semiconducting Materials and Devices* (pp. 549–590). Springer US. https://doi.org/10.1007/978-1-4757-9900-2_14 (accessed on 30th July 2025).
59. Grabowski, Z. R., Rotkiewicz, K., & Rettig, W. (2003). Structural changes accompanying intramolecular electron transfer: Focus on twisted intramolecular charge-transfer states and structures. *Chemical Reviews, 103*(10), 3899–4032. https://doi.org/10.1021/cr9407451 (accessed on 30th July 2025).
60. Grant, D. M., Mayne, C. L., Liu, F., & Xiang, T. X. (2002). Spin-lattice relaxation of coupled nuclear spins with applications to molecular motion in liquids (world) [Research-article]. *ACS Publications; American Chemical Society*. https://doi.org/10.1021/cr00007a015 (accessed on 30th July 2025).
61. Greve, C., Preketes, N. K., Costard, R., Koeppe, B., Fidder, H., Nibbering, E. T. J., Temps, F., Mukamel, S., &Elsaesser, T. (2012). N–H stretching modes of adenosine monomer in solution studied by ultrafast nonlinear infrared spectroscopy and ab initio calculations. *The Journal of Physical*

Chemistry A, 116(29), 7636–7644. https://doi.org/10.1021/jp303864m (accessed on 30th July 2025).

62. Grillet, N., Manchon, D., Bertorelle, F., Bonnet, C., Broyer, M., Cottancin, E., Lermé, J., Hillenkamp, M., & Pellarin, M. (2011). Plasmon coupling in silver nanocube dimers: Resonance splitting induced by edge rounding. *ACS Nano, 5*(12), 9450–9462. https://doi.org/10.1021/nn2041329 (accessed on 30th July 2025).

63. Gruenbaum, S. M., Tainter, C. J., Shi, L., Ni, Y., & Skinner, J. L. (2013). Robustness of frequency, transition dipole, and coupling maps for water vibrational spectroscopy. *Journal of Chemical Theory and Computation, 9*(7), 3109–3117. https://doi.org/10.1021/ct400292q (accessed on 30th July 2025).

64. Grupen, C. (Ed.). (2005). The standard model of elementary particles. In *Astroparticle Physics* (pp. 21–33). Springer. https://doi.org/10.1007/3-540-27670-X_2 (accessed on 30th July 2025).

65. Gu, Y., Zhou, Y., Tang, H., Rothe, E. W., & Reck, G. P. (2000). Pressure dependence of vibrational Raman scattering of narrow-band, 248-nm, laser light by H_2, N_2, O_2, CO_2, CH_4, C_2H_6, and C_3H_8 as high as 97 bar. *Applied Physics B, 71*(6), 865–871. https://doi.org/10.1007/s003400000412 (accessed on 30th July 2025).

66. Gunawardane, R. P., &Arumainayagam, C. R. (2006). Auger electron spectroscopy. In D. R. Vij (Ed.), *Handbook of Applied Solid State Spectroscopy* (pp. 451–483). Springer US. https://doi.org/10.1007/0-387-37590-2_10 (accessed on 30th July 2025).

67. Gupta, P., Das, S. S., & Singh, N. B. (2023). Nuclear magnetic resonance spectroscopy. In *Spectroscopy*. Jenny Stanford Publishing.

68. Gupta, T. K. (2013). Nuclear radiation, ionization, and radioactivity. In T. K. Gupta (Ed.), *Radiation, Ionization, And Detection in Nuclear Medicine* (pp. 1–57). Springer. https://doi.org/10.1007/978-3-642-34076-5_1 (accessed on 30th July 2025).

69. Gys, T. (2015). Micro-channel plates and vacuum detectors. *Nuclear Instruments and Methods in Physics Research Section A: Accelerators, Spectrometers, Detectors and Associated Equipment, 787*, 254–260. https://doi.org/10.1016/j.nima.2014.12.044 (accessed on 30th July 2025).

70. Harbecke, B., Heinz, B., Offermann, V., &Theiß, W. (1996). Far-infrared spectroscopy. In G. Bauer & W. Richter (Eds.), *Optical Characterization of Epitaxial Semiconductor Layers* (pp. 203–286). Springer. https://doi.org/10.1007/978-3-642-79678-4_5 (accessed on 30th July 2025).

71. Hemida, M., Haddad, P. R., Lam, S. C., Coates, L. J., Riley, F., Diaz,

A., Gooley, A. A., Wirth, H. J., Guinness, S., Sekulic, S., & Paull, B. (2021). Small footprint liquid chromatography-mass spectrometry for pharmaceutical reaction monitoring and automated process analysis. *Journal of Chromatography A, 1656*, 462545. https://doi.org/10.1016/j.chroma.2021.462545 (accessed on 30th July 2025).

72. Hochstrasser, R. M. (2002). Analytical and structural aspects of vibronic interactions in the ultraviolet spectra of organic molecules (world) [Research-article]. *ACS Publications; American Chemical Society*. https://doi.org/10.1021/ar50009a002 (accessed on 30th July 2025).

73. Huddart, B. M., Hernández-Melián, A., Hicken, T. J., Gomilšek, M., Hawkhead, Z., Clark, S. J., Pratt, F. L., & Lancaster, T. (2022). MuFinder: A program to determine and analyses muon stopping sites. *Computer Physics Communications, 280*, 108488. https://doi.org/10.1016/j.cpc.2022.108488 (accessed on 30th July 2025).

74. Huerta-Viga, A., Shaw, D. J., &Woutersen, S. (2010). pH dependence of the conformation of small peptides investigated with two-dimensional vibrational spectroscopy. *The Journal of Physical Chemistry B, 114*(46), 15212–15220. https://doi.org/10.1021/jp105133r (accessed on 30th July 2025).

75. Ibrahim, M. A., &Abdellatif, A. A. H. (2021). Applications of nanopharmaceuticals in delivery and targeting. In V. K. Yata, S. Ranjan, N. Dasgupta, & E. Lichtfouse (Eds.), *Nanopharmaceuticals: Principles and Applications Vol. 1* (pp. 73–114). Springer International Publishing. https://doi.org/10.1007/978-3-030-44925-4_2 (accessed on 30th July 2025).

76. Itoh, T. (2012). Fluorescence and phosphorescence from higher excited states of organic molecules. *Chemical Reviews, 112*(8), 4541–4568. https://doi.org/10.1021/cr200166m (accessed on 30th July 2025).

77. Jansen, T. l. C., &Knoester, J. (2009). Waiting time dynamics in two-dimensional infrared spectroscopy. *Accounts of Chemical Research, 42*(9), 1405–1411. https://doi.org/10.1021/ar900025a (accessed on 30th July 2025).

78. Jayasooriya, U. A., Clayden, N. J., Steytler, D. C., Oganesyan, V. S., Peck, J. N. T., Khasanov, R., Scheuermann, R., &Stoykov, A. (2016). Rate of molecular transfer of allyl alcohol across an AOT surfactant layer using muon spin spectroscopy. *Langmuir, 32*(3), 664–672. https://doi.org/10.1021/acs.langmuir.5b03482 (accessed on 30th July 2025).

79. Jegerlehner, F., &Nyffeler, A. (2009). The muon *g–2*. *Physics Reports, 477*(1), 1–110. https://doi.org/10.1016/j.physrep.2009.04.003 (accessed on 30th July 2025).

80. Jellison, G. E., & Joshi, P. C. (2018). Crystalline silicon solar cells. In H. Fujiwara & R. W. Collins (Eds.), *Spectroscopic Ellipsometry for Photovoltaics: Volume 1: Fundamental Principles and Solar Cell characterization* (pp. 201–225). Springer International Publishing. https://doi.org/10.1007/978-3-319-75377-5_8 (accessed on 30th July 2025).

81. Jones, R. R., Hooper, D. C., Zhang, L., Wolverson, D., &Valev, V. K. (2019). Raman techniques: Fundamentals and frontiers. *Nanoscale Research Letters, 14*(1), 231. https://doi.org/10.1186/s11671-019-3039-2 (accessed on 30th July 2025).

82. Kadlečíková, M., Vančo, Ľ., Breza, J., Mikolášek, M., Hušeková, K., Fröhlich, K., Procel, P., Zeman, M., & Isabella, O. (2022). Raman spectroscopy of silicon with nanostructured surface. *Optik, 257*, 168869. https://doi.org/10.1016/j.ijleo.2022.168869 (accessed on 30th July 2025).

83. Karlsson, E. (1982). The use of positive muons in metal physics. *Physics Reports, 82*(5), 271–338. https://doi.org/10.1016/0370-1573(82)90155-7 (accessed on 30th July 2025).

84. Kawanishi, T. (2022). Role of optical modulation in photonic networks. In T. Kawanishi (Ed.), *Electro-Optic Modulation for Photonic Networks: Precise and High-Speed Control of Lightwaves* (pp. 7–24). Springer International Publishing. https://doi.org/10.1007/978-3-030-86720-1_2 (accessed on 30th July 2025).

85. Khan, F., Ekbote, A., Singh, G., &Misra, R. (2022). Mechanochromicluminogens with hypsochromically shifted emission switching property: Recent advances and perspectives. *Journal of Materials Chemistry C, 10*(13), 5024–5064. https://doi.org/10.1039/D1TC06140B (accessed on 30th July 2025).

86. Koczoń, P., Hołaj-Krzak, J. T., Palani, B. K., Bolewski, T., Dąbrowski, J., Bartyzel, B. J., &Gruczyńska-Sękowska, E. (2023). The analytical possibilities of FT-IR spectroscopy powered by vibrating molecules. *International Journal of Molecular Sciences, 24*(2), Article 2. https://doi.org/10.3390/ijms24021013 (accessed on 30th July 2025).

87. Kornyshev, A. A., Kuznetsov, A. M., Spohr, E., &Ulstrup, J. (2003). Kinetics of proton transport in water. *The Journal of Physical Chemistry B, 107*(15), 3351–3366. https://doi.org/10.1021/jp020857d (accessed on 30th July 2025).

88. Krapp, A., Bickelhaupt, F. M., &Frenking, G. (2006). Orbital overlap and chemical bonding. *Chemistry–A European Journal, 12*(36), 9196–9216. https://doi.org/10.1002/chem.200600564 (accessed on 30th July 2025).

89. Kuzmany, H. (2009). Light scattering spectroscopy. In H. Kuzmany (Ed.),

Solid-State Spectroscopy: An Introduction (pp. 183–215). Springer. https://doi.org/10.1007/978-3-642-01479-6_9 (accessed on 30th July 2025).

90. Laane, J., &Ocola, E. J. (2012). Applications of symmetry and group theory for the investigation of molecular vibrations. *Acta ApplicandaeMathematicae, 118*(1), 3–24. https://doi.org/10.1007/s10440-012-9675-5 (accessed on 30th July 2025).

91. Lasorne, B., Worth, G. A., & Robb, M. A. (2011). Excited-state dynamics. *WIREs Computational Molecular Science, 1*(3), 460–475. https://doi.org/10.1002/wcms.26 (accessed on 30th July 2025).

92. Leenheer, J. A., Wershaw, R. L., Brown, G. K., & Reddy, M. M. (2003). Characterization and diagenesis of strong-acid carboxyl groups in humic substances. *Applied Geochemistry, 18*(3), 471–482. https://doi.org/10.1016/S0883-2927(02)00100-2 (accessed on 30th July 2025).

93. Leutwyler, S., & Boesiger, J. (2002). Rare-gas solvent clusters: Spectra, structures, and order-disorder transitions (world) [Research-article]. *ACS Publications; American Chemical Society*. https://doi.org/10.1021/cr00101a004 (accessed on 30th July 2025).

94. Liu, L., Zhu, Y. P., Su, M., & Yuan, Z. Y. (2015). Metal-free carbonaceous materials as promising heterogeneous catalysts. *ChemCatChem, 7*(18), 2765–2787. https://doi.org/10.1002/cctc.201500350 (accessed on 30th July 2025).

95. Lockwood, D. J. (2023). Rayleigh and Mie scattering. In R. Shamey (Ed.), *Encyclopedia of Color Science and Technology* (pp. 1361–1372). Springer International Publishing. https://doi.org/10.1007/978-3-030-89862-5_218 (accessed on 30th July 2025).

96. Lodewyk, M. W., Siebert, M. R., &Tantillo, D. J. (2012). Computational prediction of ^{1}H and ^{13}C chemical shifts: A useful tool for natural product, mechanistic, and synthetic organic chemistry. *Chemical Reviews, 112*(3), 1839–1862. https://doi.org/10.1021/cr200106v (accessed on 30th July 2025).

97. Loiko, P. A., Serres, J. M., Mateos, X., Liu, J., Zhang, H., Yasukevich, A. S., Yumashev, K. V., Petrov, V., Griebner, U., Aguiló, M., & Díaz, F. (2016). Passive Q-switching of Yb bulk lasers by a graphene saturable absorber. *Applied Physics B, 122*(4), 105. https://doi.org/10.1007/s00340-016-6384-1 (accessed on 30th July 2025).

98. Lu, X., Zhang, C., Ulrich, N., Xiao, M., Ma, Y. H., & Chen, Z. (2017). Studying polymer surfaces and interfaces with sum frequency generation vibrational spectroscopy. *Analytical Chemistry, 89*(1), 466–489. https://doi.org/10.1021/acs.analchem.6b04320 (accessed on 30th July 2025).

99. Mallick, P. K. (2023). Nuclear magnetic resonance (NMR), electron spin or paramagnetic resonance (ESR/EPR) and nuclear quadrupole resonance (NQR) spectroscopy. In P. K. Mallick (Ed.), *Fundamentals of Molecular Spectroscopy* (pp. 347–415). Springer Nature. https://doi.org/10.1007/978-981-99-0791-5_9 (accessed on 30th July 2025).
100. Martins, J. C., Biesemans, M., & Willem, R. (2000). Tin NMR based methodologies and their use in structural tin chemistry. *Progress in Nuclear Magnetic Resonance Spectroscopy, 36*(4), 271–322. https://doi.org/10.1016/S0079-6565(00)00019-4 (accessed on 30th July 2025).
101. Masuch, S., Gümbel, P., Kaden, N., &Dröder, K. (2023). Applications and development of X-ray inspection techniques in battery cell production. *Processes, 11*(1), Article 1. https://doi.org/10.3390/pr11010010 (accessed on 30th July 2025).
102. Mayer, K. M., & Hafner, J. H. (2011). Localized surface plasmon resonance sensors. *Chemical Reviews, 111*(6), 3828–3857. https://doi.org/10.1021/cr100313v (accessed on 30th July 2025).
103. McLafferty, F. W. (2011). A century of progress in molecular mass spectrometry. *Annual Review of Analytical Chemistry, 4*, 1–22. https://doi.org/10.1146/annurev-anchem-061010-114018 (accessed on 30th July 2025).
104. Michalik, M., Szymańczyk, J., Stajnke, M., Ochrymiuk, T., &Cenian, A. (2021). Medical applications of diode lasers: Pulsed versus continuous wave (cw) regime. *Micromachines, 12*(6), Article 6. https://doi.org/10.3390/mi12060710 (accessed on 30th July 2025).
105. Minhas, A. S., & Oliver, R. (2022). Magnetic resonance imaging basics. In R. Sadleir & A. S. Minhas (Eds.), *Electrical Properties of Tissues: Quantitative Magnetic Resonance Mapping* (pp. 47–82). Springer International Publishing. https://doi.org/10.1007/978-3-031-03873-0_3 (accessed on 30th July 2025).
106. Misra, A. P., &Brodin, G. (2022). Wave-particle interactions in quantum plasmas. *Reviews of Modern Plasma Physics, 6*(1), 5. https://doi.org/10.1007/s41614-022-00063-7 (accessed on 30th July 2025).
107. Mo, Y., Danovich, D., & Shaik, S. (2022). The roles of charge transfer and polarization in non-covalent interactions: A perspective from ab initio valence bond methods. *Journal of Molecular Modeling, 28*(9), 274. https://doi.org/10.1007/s00894-022-05187-8 (accessed on 30th July 2025).
108. Möbius, K., &Savitsky, A. (2023). High-field/high-frequency EPR spectroscopy in protein research: Principles and examples. *Applied Magnetic Resonance, 54*(2), 207–287. https://doi.org/10.1007/s00723-022-01511-w (accessed on 30th July 2025).

109. Moghaddam, S., & O'Leary, S. K. (2020). A Sellmeier extended empirical model for the spectral dependence of the refractive index applied to the case of thin-film silicon and some of its more common alloys. *Journal of Materials Science: Materials in Electronics, 31*(1), 212–225. https://doi.org/10.1007/s10854-019-02457-9 (accessed on 30th July 2025).

110. Mohammad-Jafarieh, P., Akbarzadeh, A., Salamat-Ahangari, R., Pourhassan-Moghaddam, M., &Jamshidi-Ghaleh, K. (2021). Solvent effect on the absorption and emission spectra of carbon dots: Evaluation of ground and excited state dipole moment. *BMC Chemistry, 15*(1), 53. https://doi.org/10.1186/s13065-021-00779-6 (accessed on 30th July 2025).

111. Monga, A., & Nandini, D. (2024). Synthetic access to thiols: A review. *Journal of Chemical Sciences, 136*(4), 67. https://doi.org/10.1007/s12039-024-02300-7 (accessed on 30th July 2025).

112. Moresi, G., &Magin, R. (2003). Miniature permanent magnet for table-top NMR. *Concepts in Magnetic Resonance Part B: Magnetic Resonance Engineering, 19B*(1), 35–43. https://doi.org/10.1002/cmr.b.10082 (accessed on 30th July 2025).

113. Motahari, H., Mousavi, S. S., &Haghighi, P. (2023). Raman, FTIR, and UV–Vis spectroscopic investigation of some oils and their hierarchical agglomerative clustering (HAC). *Food Analytical Methods, 16*(7), 1237–1251. https://doi.org/10.1007/s12161-023-02481-w (accessed on 30th July 2025).

114. Mukherji, P., & Mukhopadhyay, A. (2018). Sir Chandrasekhara Venkata Raman (1888–1970). In P. Mukherji & A. Mukhopadhyay (Eds.), *History of the Calcutta School of Physical Sciences* (pp. 21–76). Springer. https://doi.org/10.1007/978-981-13-0295-4_2 (accessed on 30th July 2025).

115. Murakami, H. (2013). Chapter six—Protein and water confined in nanometer-scale reverse micelles studied by near infrared, terahertz, and ultrafast visible spectroscopies. In C. Z. Christov (Ed.), *Advances in Protein Chemistry and Structural Biology* (Vol. 93, pp. 183–211). Academic Press. https://doi.org/10.1016/B978-0-12-416596-0.00006-3 (accessed on 30th July 2025).

116. Myers, J. A., Curtis, B. S., & Curtis, W. R. (2013). Improving accuracy of cell and chromophore concentration measurements using optical density. *BMC Biophysics, 6*(1), 4. https://doi.org/10.1186/2046-1682-6-4 (accessed on 30th July 2025).

117. Nazarski, R. B. (2023). On the use of deuterated organic solvents without TMS to report $^{1}H/^{13}C$ NMR spectral data of organic compounds: Current

state of the method, its pitfalls and benefits, and related issues. *Molecules, 28*(11), Article 11. https://doi.org/10.3390/molecules28114369 (accessed on 30th July 2025).

118. Nefedov, V. I., Yarzhemsky, V. G., &Trzhaskovskaya, M. B. (2008). The influence of relaxation and nondipole effects on the intensity of X-ray photoelectron spectra. *Bulletin of the Russian Academy of Sciences: Physics, 72*(4), 423–428. https://doi.org/10.3103/S1062873808040011 (accessed on 30th July 2025).
119. Nestler, T., Fedotov, S., Leisegang, T., & Meyer, D. C. (2019). Towards Al^{3+} mobility in crystalline solids: Critical review and analysis. *Critical Reviews in Solid State and Materials Sciences, 44*(4), 298–323. https://doi.org/10.1080/10408436.2018.1490248 (accessed on 30th July 2025).
120. Nimerovsky, E., Movellan, K. T., Zhang, X. C., Forster, M. C., Najbauer, E., Xue, K., Dervişoğlu, R., Giller, K., Griesinger, C., Becker, S., & Andreas, L. B. (2021). Proton detected solid-state NMR of membrane proteins at 28 Tesla (1.2 GHz) and 100 kHz magic-angle spinning. *Biomolecules, 11*(5), Article 5. https://doi.org/10.3390/biom11050752 (accessed on 30th July 2025).
121. Novák, J. (1974). Quantitative analysis by gas chromatography. In *Advances in Chromatography*. CRC Press.
122. Panajotov, K., &Prati, F. (2013). Polarization dynamics of VCSELs. In R. Michalzik (Ed.), *VCSELs: Fundamentals, Technology and Applications of Vertical-Cavity Surface-Emitting Lasers* (pp. 181–231). Springer. https://doi.org/10.1007/978-3-642-24986-0_6 (accessed on 30th July 2025).
123. Parks, G. K. (2018). Basic equations and concepts. In G. K. Parks (Ed.), *Characterizing Space Plasmas: A Data Driven Approach* (pp. 1–43). Springer International Publishing. https://doi.org/10.1007/978-3-319-90041-4_1 (accessed on 30th July 2025).
124. Parley, F. J. M. (1975). The anomalous magnetic moment of the muon. *Contemporary Physics, 16*(5), 413–441. https://doi.org/10.1080/00107517508210823 (accessed on 30th July 2025).
125. Pastor, K., Ilić, M., Vujić, D., Ačanski, M., Kravić, S., Stojanović, Z., &Đurović, A. (2023). Gas chromatography and mass spectrometry: The technique. In K. Pastor (Ed.), *Emerging Food Authentication Methodologies Using GC/MS* (pp. 3–31). Springer International Publishing. https://doi.org/10.1007/978-3-031-30288-6_1 (accessed on 30th July 2025).
126. Peacock, N. T. (1998). Chapter 4.2—Flange and component systems. In D. M. Hoffman, B. Singh, & J. H. Thomas (Eds.), *Handbook of Vacuum Science and Technology* (pp. 409–432). Academic Press. https://doi.

org/10.1016/B978-012352065-4/50063-8 (accessed on 30th July 2025).

127. Pfaender, H. G. (1996). Special glasses and their uses. In H. G. Pfaender (Ed.), *Schott Guide to Glass* (pp. 121–189). Springer Netherlands. https://doi.org/10.1007/978-94-011-0517-0_6 (accessed on 30th July 2025).

128. Philippidis, A., Mikallou, A., & Anglos, D. (2021). Determining optimum irradiation conditions for the analysis of vermilion by Raman spectroscopy. *The European Physical Journal Plus, 136*(11), 1194. https://doi.org/10.1140/epjp/s13360-021-02186-1 (accessed on 30th July 2025).

129. Pollnau, M., & Eichhorn, M. (2020). Spectral coherence, Part I: Passive-resonator linewidth, fundamental laser linewidth, and Schawlow-Townes approximation. *Progress in Quantum Electronics, 72*, 100255. https://doi.org/10.1016/j.pquantelec.2020.100255 (accessed on 30th July 2025).

130. Powell, C. J., & Jablonski, A. (2009). Surface sensitivity of X-ray photoelectron spectroscopy. *Nuclear Instruments and Methods in Physics Research Section A: Accelerators, Spectrometers, Detectors and Associated Equipment, 601*(1), 54–65. https://doi.org/10.1016/j.nima.2008.12.103 (accessed on 30th July 2025).

131. Qi, X., Liu, P., Qu, H., Liu, C., Bao, G., Wang, X., Liu, Y., Xin, Y., Cao, H., Chen, J., Xiao, A., Zhao, Y., Xue, B., Xu, W., Shu, R., & Ling, Z. (2025). Design and development of a stand-off Raman brassboard (SDU-RRS) for the spectroscopic study of planetary materials. *Spectrochimica Acta Part A: Molecular and Biomolecular Spectroscopy, 325*, 125026. https://doi.org/10.1016/j.saa.2024.125026 (accessed on 30th July 2025).

132. Ramirez,F.,Prasad,V.A.V.,&Marecek,J.F.(2002).Crystallinesix-coordinate phosphorus compounds derived from spiropentaoxyphosphoranes (world). *ACS Publications; American Chemical Society*. https://doi.org/10.1021/ja00830a016 (accessed on 30th July 2025).

133. Raut, V., Wani, R. R., Chaudhari, H. K., & Das, D. (2021). Solvent-free one pot synthesis of 1,2-dihydroquinolines from anilines and acetone catalyzed by MOF-199. *Results in Chemistry, 3*, 100097. https://doi.org/10.1016/j.rechem.2021.100097 (accessed on 30th July 2025).

134. Rehr, J. J., &Ankudinov, A. L. (2005). Progress in the theory and interpretation of XANES. *Coordination Chemistry Reviews, 249*(1), 131–140. https://doi.org/10.1016/j.ccr.2004.02.014 (accessed on 30th July 2025).

135. Reichert, D. (2005). NMR studies of dynamic processes in organic solids. In *Annual Reports on NMR Spectroscopy* (Vol. 55, pp. 159–203). Academic Press. https://doi.org/10.1016/S0066-4103(04)55003-2 (accessed on 30th July 2025).

136. Ritgen, U. (2025). Nuclear magnetic resonance spectroscopy (NMR). In U. Ritgen (Ed.), *Analytical Chemistry II* (pp. 27–78). Springer. https://doi.org/10.1007/978-3-662-68710-9_3 (accessed on 30th July 2025).

137. Robertson, A. D., & Shaffer, P. S. (2014). "Combustion always produces carbon dioxide and water": A discussion of university chemistry students' use of rules in place of principles. *Chemistry Education Research and Practice, 15*(4), 763–776. https://doi.org/10.1039/C4RP00089G (accessed on 30th July 2025).

138. Rohrer, M., Brügmann, O., Kinzer, B., &Prisner, T. F. (2001). High-field/high-frequency EPR spectrometer operating in pulsed and continuous-wave mode at 180 GHz. *Applied Magnetic Resonance, 21*(3), 257–274. https://doi.org/10.1007/BF03162406 (accessed on 30th July 2025).

139. Romeike, B. F. M., Meyer, T., Reichart, R., Kalff, R., Petersen, I., Dietzek, B., & Popp, J. (2015). Coherent anti-Stokes Raman scattering and two photon excited fluorescence for neurosurgery. *Clinical Neurology and Neurosurgery, 131*, 42–46. https://doi.org/10.1016/j.clineuro.2015.01.022 (accessed on 30th July 2025).

140. Rubtsov, I. V. (2009). Relaxation-assisted two-dimensional infrared (RA 2DIR) method: Accessing distances over 10 Å and measuring bond connectivity patterns. *Accounts of Chemical Research, 42*(9), 1385–1394. https://doi.org/10.1021/ar900008p (accessed on 30th July 2025).

141. Rzhevskii, A. (2024). Raman spectroscopy for the chemical analysis of food: Basic principles, instrumentation, and techniques. In *Raman Spectroscopy in The Food Industry*. CRC Press.

142. Sadeghi, K. (2023). Lagrangian potential functions of surface forces and their role in fluid mechanics. *Journal of the Brazilian Society of Mechanical Sciences and Engineering, 45*(8), 417. https://doi.org/10.1007/s40430-023-04332-5 (accessed on 30th July 2025).

143. Sanghera, J., Kim, W., Villalobos, G., Shaw, B., Baker, C., Frantz, J., Sadowski, B., & Aggarwal, I. (2012). Ceramic laser materials. *Materials, 5*(2), Article 2. https://doi.org/10.3390/ma5020258 (accessed on 30th July 2025).

144. Schlueter, J. A., Park, H., Halder, G. J., Armand, W. R., Dunmars, C., Chapman, K. W., Manson, J. L., Singleton, J., McDonald, R., Plonczak, A., Kang, J., Lee, C., Whangbo, M. H., Lancaster, T., Steele, A. J., Franke, I., Wright, J. D., Blundell, S. J., Pratt, F. L., &Landee, C. P. (2012). Importance of halogen···halogen contacts for the structural and magnetic properties of CuX_2(pyrazine-N,N′-dioxide)$(H_2O)_2$ (X = Cl and Br). *Inorganic Chemistry, 51*(4), 2121–2129. https://doi.org/10.1021/ic201924q (accessed on 30th July 2025).

145. Senanayake, S. D., Waterhouse, G. I. N., Chan, A. S. Y., Madey, T. E., Mullins, D. R., &Idriss, H. (2007). The reactions of water vapor on the surfaces of stoichiometric and reduced uranium dioxide: A high resolution XPS study. *Catalysis Today, 120*(2), 151–157. https://doi.org/10.1016/j.cattod.2006.07.040 (accessed on 30th July 2025).

146. Shao, Y., Gan, Z., Epifanovsky, E., Gilbert, A. T. B., Wormit, M., Kussmann, J., Lange, A. W., Behn, A., Deng, J., Feng, X., Ghosh, D., Goldey, M., Horn, P. R., Jacobson, L. D., Kaliman, I., Khaliullin, R. Z., Kuś, T., Landau, A., Liu, J., & Head-Gordon, M. (2015). Advances in molecular quantum chemistry contained in the Q-Chem 4 program package. *Molecular Physics, 113*(2), 184–215. https://doi.org/10.1080/00268976.2014.952696 (accessed on 30th July 2025).

147. Sinha, S., Jeyaseelan, C., Singh, G., Munjal, T., & Paul, D. (2023). Chapter 8—Spectroscopy—Principle, types, and applications. In A. K. Bhatt, R. K. Bhatia, & T. C. Bhalla (Eds.), *Basic Biotechniques for Bioprocess and Bioentrepreneurship* (pp. 145–164). Academic Press. https://doi.org/10.1016/B978-0-12-816109-8.00008-8 (accessed on 30th July 2025).

148. Smith, A. W., Lessing, J., Ganim, Z., Peng, C. S., Tokmakoff, A., Roy, S., Jansen, T. L. C., &Knoester, J. (2010). Melting of a β-hairpin peptide using isotope-edited 2D IR spectroscopy and simulations. *The Journal of Physical Chemistry B, 114*(34), 10913–10924. https://doi.org/10.1021/jp104017h (accessed on 30th July 2025).

149. Sobczyk, L., Grabowski, S. J., & Krygowski, T. M. (2005). Interrelation between H-bond and π-electron delocalization. *Chemical Reviews, 105*(10), 3513–3560. https://doi.org/10.1021/cr030083c (accessed on 30th July 2025).

150. Song, F., & Lin, X. (2024). Physical basis of laser cleaning. In F. Song & X. Lin (Eds.), *Laser Cleaning: Fundamentals and Applications* (pp. 93–111). Springer Nature. https://doi.org/10.1007/978-981-97-6631-4_3 (accessed on 30th July 2025).

151. Srividya, J., Sivamadhavi, V., Vinitha, G., &Anbalagan, G. (2022). Crystal structure, molecular structural analysis, optical absorption and thermal analysis of 2-amino-6-methyl pyridine-1-ium 2-carboxy-6-nitro benzoate–A third harmonic nonlinear optical organic single crystal. *Chinese Journal of Physics, 78*, 409–434. https://doi.org/10.1016/j.cjph.2022.07.004 (accessed on 30th July 2025).

152. Stanisz, G. J., Odrobina, E. E., Pun, J., Escaravage, M., Graham, S. J., Bronskill, M. J., &Henkelman, R. M. (2005). T_1, T_2 relaxation and magnetization transfer in tissue at 3T. *Magnetic Resonance in Medicine,*

54(3), 507–512. https://doi.org/10.1002/mrm.20605 (accessed on 30th July 2025).

153. Stenholm, S. (1973). Quantum theory of electromagnetic fields interacting with atoms and molecules. *Physics Reports, 6*(1), 1–121. https://doi.org/10.1016/0370-1573(73)90011-2 (accessed on 30th July 2025).
154. Stich, M. I. J., Fischer, L. H., &Wolfbeis, O. S. (2010). Multiple fluorescent chemical sensing and imaging. *Chemical Society Reviews, 39*(8), 3102–3114. https://doi.org/10.1039/B909635N (accessed on 30th July 2025).
155. Suga, S., Sekiyama, A., & Tusche, C. (2021). Instrumentation and methodology. In S. Suga, A. Sekiyama, & C. Tusche (Eds.), *Photoelectron Spectroscopy: Bulk and Surface Electronic Structures* (pp. 51–109). Springer International Publishing. https://doi.org/10.1007/978-3-030-64073-6_3 (accessed on 30th July 2025).
156. Sun, J., Wang, W., & Yue, Q. (2016). Review on microwave-matter interaction fundamentals and efficient microwave-associated heating strategies. *Materials, 9*(4), Article 4. https://doi.org/10.3390/ma9040231 (accessed on 30th July 2025).
157. Taylor, P. C., Baugher, J. F., &Kriz, H. M. (2002). Magnetic resonance spectra in polycrystalline solids (world) [Research-article]. *ACS Publications; American Chemical Society*. https://doi.org/10.1021/cr60294a003 (accessed on 30th July 2025).
158. Tian, Z. Q., Ren, B., & Wu, D. Y. (2002). Surface-enhanced Raman scattering: From noble to transition metals and from rough surfaces to ordered nanostructures. *The Journal of Physical Chemistry B, 106*(37), 9463–9483. https://doi.org/10.1021/jp0257449 (accessed on 30th July 2025).
159. Tran, C. D. (2012). Acousto-optic devices (world) [Research-article]. *ACS Publications; American Chemical Society*. https://doi.org/10.1021/ac00044a722 (accessed on 30th July 2025).
160. Vargas Hernández, C. (2025a). Optical parameters. In C. Vargas Hernández (Ed.), *Introduction to Raman Spectroscopy and Its Applications* (pp. 3–33). Springer Nature Switzerland. https://doi.org/10.1007/978-3-031-77551-2_1 (accessed on 30th July 2025).
161. Vargas Hernández, C. (2025b). Raman spectra of some materials. In C. Vargas Hernández (Ed.), *Introduction to Raman Spectroscopy and Its Applications* (pp. 211–240). Springer Nature Switzerland. https://doi.org/10.1007/978-3-031-77551-2_11 (accessed on 30th July 2025).
162. Verma, H. R. (Ed.). (2007). X-ray photoelectron spectroscopy. In *Atomic and Nuclear Analytical Methods: XRF, Mössbauer, XPS, NAA and*

B63Ion-Beam Spectroscopic Techniques (pp. 213–241). Springer. https://doi.org/10.1007/978-3-540-30279-7_5 (accessed on 30th July 2025).

163. Verma, V. K. (2023). Electronic spectroscopy. In *Spectroscopy*. Jenny Stanford Publishing.

164. Wagstaff, J. L., Taylor, S. L., & Howard, M. J. (2013). Recent developments and applications of saturation transfer difference nuclear magnetic resonance (STD NMR) spectroscopy. *Molecular BioSystems, 9*(4), 571–577. https://doi.org/10.1039/C2MB25395J (accessed on 30th July 2025).

165. Waluk, J. (2006). Ground- and excited-state tautomerism in porphycenes. *Accounts of Chemical Research, 39*(12), 945–952. https://doi.org/10.1021/ar0680365 (accessed on 30th July 2025).

166. Wang, Y., Griffin, P., Jin, K., Fogel, M. L., Steele, A., & Cody, G. D. (2013). Tracing H isotope effects in the dynamic metabolic network using multi-nuclear (^{1}H, ^{2}H and ^{13}C) solid state NMR and GC–MS. *Organic Geochemistry, 57*, 84–94. https://doi.org/10.1016/j.orggeochem.2013.01.013 (accessed on 30th July 2025).

167. Wehry, E. L. (1990). Effects of molecular structure on fluorescence and phosphorescence. In *Practical Fluorescence, Second Edition* (2nd ed.). CRC Press.

168. Wei, D., Chen, S., & Liu, Q. (2015). Review of fluorescence suppression techniques in Raman spectroscopy. *Applied Spectroscopy Reviews, 50*(5), 387–406. https://doi.org/10.1080/05704928.2014.999936 (accessed on 30th July 2025).

169. Wenzel, T. J., & Chisholm, C. D. (2011). Using NMR spectroscopic methods to determine enantiomeric purity and assign absolute stereochemistry. *Progress in Nuclear Magnetic Resonance Spectroscopy, 59*(1), 1–63. https://doi.org/10.1016/j.pnmrs.2010.07.003 (accessed on 30th July 2025).

170. Weyer, L. G. (1985). Near-infrared spectroscopy of organic substances. *Applied Spectroscopy Reviews, 21*(1, 2), 1–43. https://doi.org/10.1080/05704928508060427 (accessed on 30th July 2025).

171. Whangbo, M. H., Koo, H. J., & Kremer, R. K. (2021). Spin exchanges between transition metal ions governed by the ligand p-orbitals in their magnetic orbitals. *Molecules, 26*(3), Article 3. https://doi.org/10.3390/molecules26030531 (accessed on 30th July 2025).

172. Wu, Y., & Zhu, W. (2013). Organic sensitizers from D–π–A to D–A–π–A: Effect of the internal electron-withdrawing units on molecular absorption, energy levels and photovoltaic performances. *Chemical Society Reviews, 42*(5), 2039–2058. https://doi.org/10.1039/C2CS35346F (accessed on 30th July 2025).

173. Yu, H., Mazzanti, C. L., Whitfield, T. W., Koeppe, R. E. I., Andersen, O. S., & Roux, B. (2010). A combined experimental and theoretical study of ion solvation in liquid N-methylacetamide. *Journal of the American Chemical Society, 132*(31), 10847–10856. https://doi.org/10.1021/ja103270w (accessed on 30th July 2025).

174. Yusuf, M. O. (2023). Bond characterization in cementitious material binders using Fourier-transform infrared spectroscopy. *Applied Sciences, 13*(5), Article 5. https://doi.org/10.3390/app13053353 (accessed on 30th July 2025).

175. Zhang, H., Zheng, F., Zhang, J., Hu, D., Minelli, M., Pei, J., &Sangiorgi, C. (2024). Contributions of Fourier-transform infrared spectroscopy technologies to the research of asphalt materials: A comprehensive review. *Fuel, 371*, 132078. https://doi.org/10.1016/j.fuel.2024.132078 (accessed on 30th July 2025).

176. Zhang, Y., Liang, F., Zhao, D., Jiang, D., Liu, Z., Zhu, J., Yang, J., & Liu, S. (2020). Hydrogen can passivate carbon impurities in Mg-doped GaN. *Nanoscale Research Letters, 15*(1), 38. https://doi.org/10.1186/s11671-020-3263-9 (accessed on 30th July 2025).

177. Zheng, A., Liu, S. B., & Deng, F. (2017). ^{31}P NMR chemical shifts of phosphorus probes as reliable and practical acidity scales for solid and liquid catalysts. *Chemical Reviews, 117*(19), 12475–12531. https://doi.org/10.1021/acs.chemrev.7b00289 (accessed on 30th July 2025).

178. Zhou, J., Yang, Y., Ren, X., &Stapf, S. (2006). Investigation of reinforcement of the modified carbon black from wasted tires by nuclear magnetic resonance. *Journal of Zhejiang University-SCIENCE A, 7*(8), 1440–1446. https://doi.org/10.1631/jzus.2006.A1440 (accessed on 30th July 2025).

179. Ziessel, R., Ulrich, G., Haefele, A., & Harriman, A. (2013). An artificial light-harvesting array constructed from multiple Bodipy dyes. *Journal of the American Chemical Society, 135*(30), 11330–11344. https://doi.org/10.1021/ja4049306 (accessed on 30th July 2025).0

INDEX

I

K

L

M